Allah's Oil

Also by I. G. Edmonds

Allah's Oil

Mideast Petroleum

by

I. G. Edmonds

THOMAS NELSON INC., PUBLISHERS
Nashville • New York

Library of Congress Cataloging in Publication Data
Edmonds, I G
    Allah's oil.
    Bibliography: p.
    Includes index.
    1. Petroleum industry and trade—Near East—History. I. Title.
HD9576.N36E35   338.2'7'2820956   76–28749
ISBN 0-8407-6504-5

1940881

Contents

Introduction
The Oil War

In 1973 the world's greatest "have" nation, the United States of America, received a stunning shock. It had become a "have-not" nation—at least as far as cheap and abundant energy is concerned.

Lines began to pile up at service stations. Gasoline was limited to ten gallons at a sale. Fuel oil was allocated. Lights were "browned out." Industries were told by the government to save energy or face the possibility of being shut down. As a result, business fell off. Unemployment rose alarmingly. A large part of the world was caught in a business recession.

While the energy crisis had been building up for several years, the immediate causes of the shortage were a boycott of oil sales to the United States and the Netherlands by Arab oil-producing countries and a staggering price raise for petroleum by all the major oil producers outside the United States. At the same time, the Arab nations reduced their oil production 25 percent so the nations they sold to would not have a surplus that they might resell to the United States.

Oil sales to the United States and the Netherlands were boycotted because these two nations had supported Israel in the 1973 Yom Kippur War between Israel, Egypt, and Syria. Iran, the other major oil producer in the Middle East, did not join in the boycott, but did go along with the tremendous price increases that threw Western industrial countries into a recession from which they recovered slowly. The price of oil quadrupled within one year.

Up to this point most Americans thought of Arabs, when they thought of them at all, as camel-riding warriors like those in *Beau Geste,* as romantic figures from *The Sheik* and *The Desert Song,* or as terrorists making trouble for Israel.

The oil crisis brought home to Americans the fact that the Arabs had suddenly become a powerful economic force in today's unsteady world. The world moves on oil, and oil fires a major portion of industry. Sixty-eight percent of this oil is controlled by Middle East countries and by Arab countries in North Africa.

Even more disturbing than the economic shock of huge increases in the price of oil is the Arabs' decision to use oil as a weapon in world politics. This oil weapon is aimed directly at the United States and is intended to break or at least lessen United States support of Israel.

During the Yom Kippur War, Arab threats to extend the oil embargo caused several nations (including Germany) to refuse to permit United States planes resupplying Israel to land at their bases.

Although the Middle East is so important that it can throw the Western world into an economic recession, few Americans know much about that part of the world. It is unfortunate, because an understanding of the Middle East oil problem is necessary before we can hope to find a solution for the many economic, political, and military problems that threaten world peace. The next and most devastating of all world wars could very well start in the Middle East if these problems are not solved.

The harsh facts are that the people of the Middle East, both Arab and non-Arab, historically have little reason to like or have pity on the Western world. We have been enemies for centuries. The trouble began when the Arabs, who had been united under the Prophet Muhammad, rode out of Arabia in 637 A.D. to conquer

Persia. In the next ninety years they carried the banner of Islam, their religion, through the rest of the Middle East, North Africa, and Spain and into France. The march of Islam was stopped by Charles Martel at the battle of Tours (145 miles from Paris) in 732.

This battle halted the Muslim move into Europe, but the Arabs were not expelled from Spain until 1492. In between were years of struggle, including the famous Crusades in which Christian kings banded together to drive the Muslims from the Holy Land.

In time the Arabs were conquered by the Turks, who also followed the Muslim religion. Turkish control of a large section of the Middle East was not broken until World War I. During this war, the Allies enlisted Arab help in fighting the Turks. The Arabs thought they were fighting for their independence, but this was denied them at the war's end.

Later the Arabs did become independent, but at the end of World War II many of them were uprooted from their homes by the United Nations in the creation of the new state of Israel, which again fanned the fires of hatred between the Arabs and the West. As a result, the Arabs have swapped their flashing scimitars for oil, which is their weapon in this new struggle.

Thus the story of Middle East oil is of interest to us because of the political and economic impact it is having upon the entire world. Thus oil threatens world peace. It is making us pay more for the things we buy, and is lowering our standard of living. It is increasing unemployment.

But the story of Middle East oil is also of interest because it is the swashbuckling story of some true pioneers. The men who opened the oil fields were no ordinary breed of men. They were indomitable.

The modern story of Middle East oil began when a retired gold-mining millionaire met a French archaeologist, who had gotten his shoes dirty while hunting ancient ruins in Persia. But the true beginning took place in a remote geological age when young seas were forming in depressions in the steaming rocks of the hot earth.

1
The Mystery of Oil

Petroleum is one of the mysteries of creation. Geologists can tell us what it is, for they have analyzed it. They can tell us where it comes from—that is, after it has been found by trial and error—but not one can tell us for sure how it got under the ground in the first place. The most widely held theory is that petroleum was created from marine life and vegetable matter, which were deposited in incredibly thick layers along the shores of prehistoric seas. These layers were later covered with sediments that turned to rock. Then pressure, heat, and possibly bacterial action turned the organic matter into the hydrocarbon substance we call petroleum, which means rock oil.

In the beginning, cosmologists, those learned people who ponder on the origin of creation in a scientific way, say that huge masses of cosmic dust floated in the emptiness of space. In time the motion of this primordial dust caused it to spiral. Eventually motion and gravity condensed it through tremendous heat into stars and planets.

According to this theory, the burning earth cooled. A crust formed over the hot surface. Heat and pressure from the unstable core caused this crust to crack, buckle, and fold, creating huge mountain ranges. Gases broke through the cracked crust to form an atmosphere.

Moisture condensed from the atmospheric gases and rained back on the hot earth. The constant beat of water eroded the rocks and created soil. The water drained into shallow seas, which boiled away to make more clouds and rain.

Heat, pressure, and churning from the unstable crust worked to create compounds from the elements in the steaming seas. In time these compounds became even more complex, and that is the stage at which life appeared. Life could have been created by divine intervention, by the application of natural forces operating under the right conditions, or by pure accident. Opinions vary widely.

The first life-form was microscopic, as fossilized rocks prove. It began some two billion years ago in the Pre-Cambrian era. The succeeding Cambrian period was the first geological division of the Paleozoic era, which extended from about 550 million years ago to the end of the Permian period, about 190 million years ago.

During the Paleozoic era much of the earth was covered with shallow seas, teeming with minute marine life. In time these marine creatures grew backbones and became fishes. On land, primitive plant life developed and eventually grew into huge fernlike trees. Other land life followed, first perhaps scorpions and finally reptiles which grew into the dinosaurs that dominated earth during the next or Mesozoic, era.

Geologists believe that oil was formed in the Paleozoic time. Marine and plant life with a hydrocarbon base were deposited in huge layers along the shores of the shallow seas. The layers were covered by sediments that became rock. The sinking and folding of the buckling earth, during violent periods of mountain-building, carried the hydrocarbon layers deep into the earth where heat, pressure, and bacterial action turned the matter into petroleum. This process probably occurred droplet by droplet.

The drops of oil seeped through the porous limestone and

sandstone with which they were associated, and were finally stopped by layers of shale and nonporous rock so that they collected in pools. Sometimes these droplets collected in domes of porous rock where they seeped through like water into a sponge. Sometimes the oil collected in traps left by the movement of earth faults. Sometimes it collected in salt domes. Whatever the place it collected, the oil was locked in by the nonporous rock above and below it.

Just as earth movements carried the original deposits deep into the earth, sometimes the folds and uplifts of the earth's crust reraised the oil deposits close to the surface. Then cracks and faults created seepages that let the trapped oil or gas escape. The famed La Brea tar pits of Los Angeles are an example of surface seepage of petroleum deposits.

Prehistoric man knew and used these seepages. Assyrians used the tar to waterproof their boats in Mesopotamia. The ancient Egyptians used pitch (natural asphalt) to coat their mummies. The American Indians used seepage oil both as a fuel and as a medicine.

Although widely used in a limited sense, oil did not become important until the nineteenth century. Salt-makers, drilling wells to find salt, often found petroleum mixed with the salt. Then in 1847 a Pittsburgh, Pennsylvania, salt-maker named Samuel M. Kier decided to get some use from the oil that kept appearing in his salt wells.

Kier knew that the Indians used petroleum for medicine. He began bottling the crude oil from his wells. This he peddled as an "old Indian remedy" under the trade name of Kier's Rock Oil. He advertised it as an infallible cure for cholera, bronchitis, liver ailments, and consumption. He recommended that sufferers take three teaspoonfuls three times a day.

Five years after Kier began selling rock oil, Abraham Gesner, a Canadian, discovered kerosene, a light oil he extracted from coal. This gave kerosene ("wax oil") its popular name of "coal oil." Kerosene was hailed as the long-sought substitute for whale oil, then in use for lamp lighting. Whale oil was expensive and smelly, and smoked badly. However, it was an improvement in lighting over the tallow candle. Kerosene became a practical source of illumination in 1853, when a German named Stohwasser invented a lamp with a glass chimney that worked well with it.

In the meantime Kier kept selling his rock oil. Having a scientific turn of mind, he wanted to know what was in it. He had a chemist analyze a sample. The report said that the crude oil (that is, oil as it comes directly from the ground) was a complex hydrocarbon compound. The chemist said that if this oil were refined, the products could be used for illumination.

Kier then built himself a still like the ones used by moonshiners to make whiskey. His output from the still was low because the only crude oil he had came from seepages in and around his salt wells, and he did not think of drilling directly for oil. Thus he missed his chance to become the first "oil king."

A young teacher named George H. Bissell was more farsighted. While visiting at Dartmouth College, he was shown a bottle of rock oil and told that it could be refined to produce a grade of kerosene much more cheaply than a similar grade could be extracted from coal. This fired Bissell's imagination. He went to Pennsylvania and leased some land that contained an oil seepage. He then formed a company called the Pennsylvania Rock Oil Company. A sample of the oil was sent to Yale University for analysis.

The report, signed by a Professor Silliman, electrified Bissell. The report said, "Your company have in their hands a raw material from which, by simple and not expensive processes, they may manufacture very valuable products. It is worthy of note that nearly the whole of the raw product may be manufactured without waste, and this solely by a well-directed process which is in practice one of the most simple in the whole of chemical science." The professor was referring to distilling, which Kier was already doing on a small scale.

Up to this point Bissell had been following in the tracks of others in the United States, Great Britain, and France. His contribution to the development of the oil industry was a decision to drill a well directly to obtain oil. He began drilling in 1857, but went bankrupt. His company was taken over by the Seneca Oil Company, headed by a group of businessmen who were interested in competing with Kier in the medicine-oil business. Seneca hired Colonel Edwin L. Drake, whose title was honorary, because Drake was merely a retired railroad conductor who had had some experience in drilling water wells.

The location for the drilling was near Titusville, Pennsylvania. Drake hired a blacksmith and his two sons, and began drilling. They used cable tools. That is, a steel bit is raised up by a derrick and permitted to fall in the hole. As it is raised and dropped repeatedly, it chips away at the rock. Drake used a small steam engine to pull the bit up.

After the hole reached thirty-nine feet in depth, water began seeping in. Drake then invented casing. That is, he drove a large pipe into the hole and drilled from inside the pipe. When he reached the end of the pipe, he drove a smaller pipe through it and kept drilling. This technique is still in use today.

Drake started drilling in February 1859. He finally struck oil at a depth of sixty-nine and a half feet on August 27, 1859. The well flowed from ten to twenty-five barrels of oil a day. The Seneca Oil Company sold it to medicine companies for twenty dollars a barrel.

After Drake showed the way, the rush to drill for oil was on, aided by the development of kerosene lamps and stoves, and then by the invention of the automobile. The story of oil from Drake to the present day is a rip-roaring adventure story of brave men, honest men, crooks, scheming politicians, and just plain nuts. However, this is not the story of oil, but of oil and the Middle East. Therefore, it is enough to say that oil became extremely valuable.

As the uses for oil expanded and Pennsylvania production dropped off, oil was found in Texas, Oklahoma, Arkansas,

Louisiana, Wyoming, California, and in smaller quantities, at other locations. At this time, except for a small amount of production in Romania, the United States supplies practically all of the world's oil. And nine tenths of American oil was controlled by a man named John Davison Rockefeller.

Rockefeller was in his early twenties when he became interested in oil in 1862. He immediately recognized the enormous potential in oil and set out to control the entire industry. Twenty years later, by 1882, his Standard Oil Company had succeeded in obtaining almost a total monopoly on all oil produced in the United States. He reached this position by buying out rival oil producers and refiners whenever he could. When that failed, he undersold them into bankruptcy. In 1882, Rockefeller formed his Standard Oil trust in order to bring all his smaller companies under one management. The new company was capitalized at $70 million. Worldwide oil consumption was then 35 million barrels a year, most of which was used for lighting. Of these barrels, Rockefeller supplied 30 million. He was able to set his own prices for oil, because he controlled so much of the market. He was doing then exactly what the Middle East oil producers are doing today.

In the same year that Rockefeller formed Standard Oil, a British naval officer, Sir John Fisher, began agitating for the Royal Navy to change from coal to oil as fuel for the ships. He complained that one fourth of the fleet was in port for recoaling all the time. "If the fleet burned oil," he insisted, "we could refuel at sea from floating tankers. This would increase the use of our vessels by 25 percent."

Senior naval commanders were as conservative then as they are today. They had no desire to make any changes. Also, they argued, Great Britain had large coal reserves, but no oil fields. It was unthinkable to convert naval vessels to oil until Great Britain controlled its own oil reserves. Otherwise, it would put the navy at the mercy of a foreign country in time of war.

This argument made sense, although Fisher continued to argue, and enterprising men, seeing the growing need for oil both for military and domestic use, began seeking new sources of it.

For British oil explorers, the most likely place to seek petroleum deposits was in the Middle East, the region of southwest Asia and northeast Africa that extends from the Mediterranean coast to the borders of Pakistan and from Turkey to Egypt. It includes Turkey, Syria, Israel, Jordan, Iraq, Iran, Lebanon, the countries of the Arabian Peninsula, and Egypt and Libya. The British government, since the conquest of India, had been extending its influence in this region to protect the supply lines between India and England.

The first person to get an oil concession in the Middle East was Baron Julius von Reuter (also founder of the famous Reuter's international agency). Reuter had to give up his concession in northern Iran (then called Persia) when the Russians objected to British operations so close to their border.

Next the British firm of Holtz and Company drilled test wells in southern Persia. These proved to be dry holes and the company went bankrupt. Reuter, in 1889, secured another concession from the Persian monarchy. It was to run sixty years, but was canceled after two years when Reuter failed to find oil.

Here the matter of Middle East oil rested until a retired millionaire mineowner with nothing to do met a French archaeologist who had dirtied his shoes in a Persian oil sump. From this meeting grew the often dramatic events that developed into the entire Middle East oil business which is upsetting today's world economy.

2
D'Arcy and the Indomitable Mr. Reynolds

According to a petroleum industry legend, William Knox D'Arcy staked the Morgan brothers with the Australian equivalent of one hundred dollars to help them prospect for gold. They immediately made the biggest strike in the history of Australia.

The story that D'Arcy got in on the fabulously rich gold mine for just a hundred dollars may or may not be true. But it is true that he helped organize the company that exploited the mine, and that he was able to retire to London in the late 1890's with a considerable fortune. However, after an exciting life in Australia, D'Arcy found his inaction in London boring. He began looking around for something to do.

In the circles D'Arcy moved in, talk was predominantly about business and finance. One of the favorite subjects was the enormous amount of money John D. Rockefeller was making in the oil business. The way Rockefeller had forced out his competitors in the United States to gain a petroleum monopoly was a matter of great admiration and envy to D'Arcy's colleagues. After all, these were the days of the "robber barons," when "cornering the market" was a prime business objective. In that way, you were able to control an industry or commodity, could set your own prices, and had no competitors to force prices down.

A reluctant United States government was being prodded into taking action against Rockefeller's oil monopoly. Various states had already passed antitrust laws to prevent companies from gaining control of an entire industry, but these state laws had been so loosely drawn that Rockefeller's oil cartel had no difficulty in finding loopholes to get around them. Rockefeller's skirmishes with these antitrust laws were of great interest to the men in D'Arcy's circle.

The petroleum industry had boomed on the basis of supplying kerosene for general illumination and heating. Then Thomas A. Edison invented the incandescent light, and it began to appear as if oil had a poor future. At the same time, its popularity as a medicine was declining.

Rockefeller did not agree that oil use would decline, and events proved him right. By 1900 it was clear that the horseless carriage with its gasoline engine would soon be using far more petroleum than lighting ever did. Also small boats were beginning to use the new gasoline engine. Germany was showing great interest in changing its coal-burning battleships to oil burners. The German interest caused Sir John Fisher, who had been agitating for the British navy to do the same thing, to renew the demands he had been making since 1882.

This increasing interest in oil caused chubby little Mr. D'Arcy to take a hard look at the future of the petroleum business in Great Britain. The country had little oil under its direct control. The Burmah Oil Company had limited production in Burma, but this amount was insufficient to feed British demands.

D'Arcy was told that the German interest in oil-burning ships would force the British navy to convert to oil also. Experts assured D'Arcy that oil would give increased speed over coal-burning ships. This knowledge made it vital to British security that the country find a major oil supply that it could control. It would be disastrous in time of war to have to depend upon Rockefeller or some other foreign source for oil to supply the Navy.

D'Arcy was certain that before long there would be a British Rockefeller, and he saw no reason why it should not be himself.

In his studies of the American oil empire, D'Arcy learned that there was no way to determine where oil was hidden in the ground. In Pennsylvania, successful oil wells were dug by drilling in places where oil seeped to the surface. However, there was no guarantee oil would be found in such places; numerous dry holes had been drilled by companies following this method. Geologists were studying the rock formations in Pennsylvania, seeking clues that would make it possible to guess more scientifically as to the location of oil pools.

So it was that D'Arcy was quite interested when he heard about the adventures of the celebrated French archaeologist, Jacques de Morgan, in his search for early ruins in the Zagros Mountains of Persia. De Morgan told of stepping into pools of tar and pitch, some of which were so deep that a man could sink in over his head, as the mammoth and saber-toothed tigers had done in California's celebrated tar pits.

Eduard Cotte, a friend of De Morgan, had been associated with Julius von Reuter in the previous attempt to get Persian concessions. Cotte and his friend, Sir Henry Drummond Wolff, had contacts with a highly placed customs official in Persia, a man of Armenian birth named Antoine Kitabgi. D'Arcy talked with De Morgan, Cotte, and Wolff, then sent a representative to Persia to work out a concession.

One of the innumerable legends that grew up around D'Arcy claims that he was not impressed by any of the arguments presented by his would-be partners. He agreed to investigate the possibilities solely because of De Morgan's name. Backing the Morgan brothers in Australia in 1882 had made D'Arcy rich. His hunch was that the Morgan name would be just as lucky in the search for oil as it had been in gold prospecting.

After considerable trouble and many payoffs, D'Arcy's representative in 1901 secured a drilling concession that included all of Persia except the provinces bordering Russia. A geologist recommended two sites to D'Arcy. One was at Chiah Surkh on the Turkish-Persian border in the northwest territory; the other, in the southwest close to the present Iraqi border.

Although D'Arcy was responsible for opening up the great Persian oil fields, he never visited the country himself. He was strictly a financier. He knew nothing of drilling or petroleum geology. All he had was faith and the example of John D. Rockefeller.

Now that he had a drilling concession, D'Arcy needed a man to do the drilling. An official of the Burmah Oil Company recommended G. B. Reynolds, who had had some experience drilling wells in Sumatra. Another of the many D'Arcy legends is that he never met Reynolds in his life. He hired the man on his friend's recommendation and sent Reynolds to Persia with a free hand to do as he thought best.

D'Arcy could not have made a better choice. Reynolds had years of experience with the British civil service in India before he went to Sumatra. He knew how to handle native workers. Moreover, he was a bulldog of a man who, once he got his teeth into a job, never let go. Nothing daunted him. Failure to him was just an incentive to try a little harder.

Reynolds was a fifty-year-old widower when D'Arcy sent him to Persia. He was stockily built and wore a walrus moustache of the type popularized by Rudyard Kipling. He habitually wore a cork sun helmet and canvas leggings. He was not a loud-voiced man, like so many drillers, but he did have an extremely sarcastic manner, which he displayed more to those above him in company rank than he did to those who worked for him.

A lesser man would have been daunted by what Reynolds found when he got to Persia. The area where he was to drill was the bleakest kind of rocky wasteland. Some scrub bushes were the only vegetation. He had to move the original site because it was too far from water. Water was an absolute necessity, both to flush the drillings out of the hole and for the boilers that created the steam to run the steam engine.

Everything used for drilling had to be packed in on mule and camel trains. This was so laboriously slow that sixteen months passed from the time Reynolds was hired until he finally started boring a hole in the grond. The boiler, steel drill bits, pumps, steam engines, crown block (a pulley system atop the derrick), and other essential drilling equipment were shipped from England to the Persian Gulf, where they were transferred to the shallow-draft Arabian sailing ships, called dhows, that took the material up the Shatt al-Arab to Basra. The Shatt al-Arab is formed of the combined Tigris and Euphrates rivers, which converge above Basra to flow as one stream into the Persian Gulf.

From Basra the heavy equipment was transported through Baghdad and up over the mountains to the drilling site. Two wide steel wheels were attached to the boiler, and mules pulled it along. Similarly, big-wheeled frames were constructed to freight the casing pipe. The natives called these ''jins.'' Smaller pieces of equipment were packed on camels and mules.

The pack trains went up the Tigris beyond Mosul and then cut across the bleak desert into the Zagros Mountains. Here they encountered rocky terrain so steep and slippery that the hooves of the mules could not get footholds to climb. They had to be roped and pulled upward by the struggling, sweating men. The camels with their padded feet did better.

The difficulties of freighting in supplies and equipment were enough to daunt the hardiest souls, but Reynolds found that his trouble was just beginning. D'Arcy's agent had made his contacts and obtained his concession from the Qajar government in Tehran. However, the Qajar government controlled only the area around Tehran, the capital. The rest of the country, while professing allegiance to the central government, was actually controlled by local chiefs, who did as they pleased.

Originally the Persian people came out of the plains of southern Russia between the years 2300 and 1500 B.C. In 559 B.C. the two major Iranian groups, the Medes and the Persians, were united under Cyrus the Great. Cyrus went on to found a great empire that extended from present-day Turkey to Egypt, including all the land between. His successors pushed the frontiers into Afghanistan and Pakistan, and came close to defeating the Greeks.

In the passing centuries the fortunes of the Persian Empire rose and fell several time. In A.D. 637 the country was conquered by the Arab Muslims in the first great victory of their *jihads,* or holy wars. The Arabs remained in control until ousted by Ismail in 1499. His dynasty in turn was superseded in 1794 by that of the Qajars.

The powerful Qajars, who had kept the Ottoman Turkish rulers from overrunning Persia, had gone to seed by the time D'Arcy made his drilling agreement with the Qajar ruler, called the Shah. The half-wild tribesmen knew that the central government could not spare troops to fight them, for the Shah needed his army in Tehran to keep from being toppled from the throne. These hillmen threatened to run Reynolds out of their territory unless they also received payments.

Chiah Surkh is located in a small bowl of land surrounded by low, barren hills. A stream, fed by melting mountain snow, cuts through the valley. Reynolds located his first well near a tar pit a half mile from this river. The distance was to keep river water from seeping into the well hole. A ditch was dug from the river to the drilling site to provide water for the boiler. The boiler furnished steam to run the engines, pumps, and electric generator.

Reynolds brought in a drilling crew of four Poles, one Canadian, and one American. Oil-company historians are unanimous in calling the crew incompetent, except for the Canadian, whose name was McNaughton. If this was so, however, one wonders why Reynolds always left the American, Hiram Rosenplaenter, in charge when he had to go to Baghdad on business.

Because of the heat, which sometimes reached 140 degrees Fahrenheit at noon during the summer, work was done mostly at night. The workmen were brought from Tehran, 150 miles away. They cleared and leveled the ground, and erected a derrick under Reynolds' direction. The derrick was used to raise the drill bit so it could be dropped into the hole. This was the cable-tool method of drilling that had been employed by Colonel Drake to drill the first oil well in Pennsylvania. The rotary drilling system used in today's operations had not yet been perfected.

On top of the derrick there was a "crown block," which was a collection of pulleys through which the cables ran. One end of the cable was wrapped around a huge wooden spindle, the "bull wheels." The other end passed through the crown block and down through the center of the derrick to suspend the bit. In operation, the driller released the catch on the bull wheels. The loosened cable, pulled by the heavy weight of the bit, plunged into the drill hole, chipping into the rock. The driller then reversed his control. The steam engine rotated the bull wheels, pulling the cable tool up the hole so it could be dropped again. When dirt and chips in the hole began to interfere with drilling, water was pumped in and the "slush" removed with a "bailer." The bailer looked like an elongated section of pipe with a trap at the end. When the bailer hit bottom, it pushed the trap open and the mixture of water, mud, and chips flowed into the bailer. When the bailer was lifted off the floor of the hole, the weight of the muck inside closed the trap. The contents were then lifted out of the hole and poured into a "slush pit." Later, when rotary drilling came into use, water was pumped into the hole continuously during the drilling operation and expended continuously into the slush pit.

In November 1902 drilling began, a slow, steady pounding away at the earth. A hole 300 feet deep had been dug into the ground by January 1, 1903. Rains came in early December, reviving the water supply, which had dwindled to a trickle. The wet weather also brought some relief from the heat, which still reached 100 degrees at midday, but the rain made life miserable by blowing into the tents and turning the ground into a bog. Reynolds had two sheet-iron huts of the type used in Australia, but they were reserved for an office and a storage place for the foreign crew's liquor.

As soon as the first well had been "spudded in" (started), Reynolds scouted for a second location. However, work pro-

Where it all started—the first well drilled by C. B. Reynolds at Chiah Surkh.

Courtesy National Iranian Oil Co.

ceeded slowly on the first hole. Part of the foreign drilling crew was replaced, but the newcomers were apparently no better than the former employees, for records show that Reynolds wrote the London office that "a more helpless crew I've seldom met."

Spring, when it finally came, was a delight. The barren hills came alive with flowers and vegetation whose seed had lain dormant waiting for water. The crew, their spirits revived, now spudded in the second well not far from the first.

Then one day the sky darkened with thick clouds of locusts. The insects settled on everything green. They stripped and devoured the vegetation so bare that a goat would have starved on the remains. The insects died by the millions. The rocks and ground were slippery with their crushed bodies.

The locust plague lasted for a miserable six days. It ended only when the swarm had eaten everything possible. Then the clouds of insects flew off to devastate a new location.

The drilling crew pushed the insect bodies into heaps and burned them. Then they went wearily back to drilling as the furnace heat of summer returned to add more misery. The steady pounding continued, slowly pushing the hole deeper into the earth. By August, the first well was down to 1,665 feet. The second struck a pocket of natural gas at seventeen feet. There was an explosion that hurled the bit out of the hole, but fortunately no one was hurt.

Drilling went on. The stream dried up again. Some days they could not drill at all, because there wasn't enough water for the boilers. They had to dam the stream and wait for the trickle of water to build up until they had enough for a day's operations. Worse yet, they were hauling drinking water from a village thirty miles away. Smallpox broke out there, and the native porters, who had been carrying water on camels, refused to go to the village.

Finally the rains came in November. This time nobody complained when it blew under the tent flaps to wet the bedding. They were glad to have it. Then, in January 1904, they ran into an exceptionally hard layer of rock. It ground off the edges of the bits, and Reynolds kept changing them so that the "tool dresser" could regrind the edges. Later in the month the drill broke through the hard crust into soft sandstone. Oil began welling up through the casing. As the drill drove deeper, crude oil under pressure shot up above the top of the derrick.

The jubilant Reynolds dispatched a message to D'Arcy in London, informing the financier that Well Number 2 was flowing oil at a rate varying from 125 to 200 barrels a day. For D'Arcy it was, as he said later, "glorious news." He had already decided that he had made a mistake and had been trying to sell part of his interest in the First Exploration Company, the organization he had set up to finance the drilling in Persia. Stock from the company had gone to pay the Persian government its concession price, and to various investors, including himself. Some was sold to the public. However, the money was quickly spent. D'Arcy then sank $750,000 of his own funds into the project. He also tried to interest other financiers into putting money in the company. Their answer was, "Come back when you strike oil."

When Reynolds reported that he had struck oil, D'Arcy thought his financial troubles were over. Unfortunately, on May 14, 1904, the Chiah Surkh well stopped flowing. It proved to be only a small pocket of oil. Then the Number 1 Well was abandoned as a dry hole.

D'Arcy's geologists recommended that the drilling crew move to southwest Persia. The prospects there appeared better. Also, they had discovered that it would take a superhuman effort to build a pipeline to Chiah Surkh if oil should be discovered there. The oil would be worthless unless they could get it to the Persian Gulf where tankers could pick it up.

D'Arcy was bankrupting himself, but he could not quit, for that would mean a total loss to him. He gave the order for Reynolds to move south, and sometime afterward he went to France to confer with representatives of the Rothschilds, the extremely wealthy banking family. Negotiations were in progress when D'Arcy received an urgent cable from London asking him to withdraw his offer to the Rothschilds.

D'Arcy readily agreed. The Rothschilds' offer would have left him with a considerable loss of money. He returned to London to meet the Hon. E. G. Pretyman, the man who had sent the cable. Pretyman was chairman of the oil committee appointed by the First Sea Lord, Admiral Sir John Fisher, the same Fisher who, as a navy commander, had been agitating for oil-burning ships since 1882. Fisher was now in a position to do something about it.

Pretyman told D'Arcy that it was definite that the German navy would convert to oil-burning ships. That made it imperative for the British navy to do the same. "Unfortunately," Pretyman told D'Arcy, "we face the same objections as when Fisher first made his suggestion. England possesses no oil except for the small fields in Burma and Assam. In the event of a major war we would have to depend upon a foreign source. We must find a British-controlled source of oil."

Pretyman went on to enumerate the advantages of oil over coal as a power source for warships. In 1923, Winston Churchill listed these in his book *The World Crisis*. Churchill said:

> First, speed. In equal ships oil gave a large excess of speed. . . . It gave 40 percent greater radius of action. . . . The use of oil made it possible for every type of vessel to have more gunpower and more speed for less size and less cost.

Pretyman went on to say that the British government did not look with favor upon D'Arcy's proposed sale to the French Rothschild interests.

"Our experts tell us that Persia has excellent prospects of delivering very rich oil fields despite your disappointment in your first venture," Pretyman said. "I can also say in all truthfulness that the British government will be a very large purchaser of your oil. This will bring a large financial return to your company."

D'Arcy frankly replied that he was nearing the end of his private resources. He did not have the money to continue drilling in Persia. Pretyman then promised to find the additional financing, if D'Arcy would hang on a little longer. D'Arcy agreed.

The promised financial help proved to be an introduction of D'Arcy to officials of the Burmah Oil Company. Burmah was interested in extending its operations into the Middle East and suggested that a syndicate of financiers be formed to carry on D'Arcy's explorations. In this way, none of those involved would lose too much if the drilling failed to find "pay sand."

D'Arcy agreed, but the Persian government objected. The Shah wanted the drilling to continue, but he saw this as an opportunity to extort more money from the new syndicate. The Persian officials claimed that their agreement had been with D'Arcy. Any change in this agreement involved a new concession and they must be paid an additional fee.

These arguments postponed the formation of the concessions syndicate until May 5, 1905. At this time, D'Arcy turned over his concession and assets in Persia to the syndicate. Burmah and Lord Strathcona, a Scots-Canadian financier, put up extra capital to continue operations. D'Arcy received stock in the syndicate and was also repaid part of his initial investment.

Reynolds was ordered to move to southwestern Persia. The new drilling area was near the Iraqi border and about 150 miles from the Persian Gulf. Company geologists selected a site at Shardin, about fifty-five miles east of the town of Ahwaz on the Karun River. As soon as the well was spudded in, Reynolds asked permission to drill a second well at Maidan-i-Naftun ("The Plain of Oil"), which was eighty miles from Shardin.

This additional well involved Reynolds in constant travel between the two sites, but by this time Reynolds, always at odds with his superiors, had become a fanatic about drilling. He had convinced himself that the company's geologists did not understand the petroleum business and were wrong in deciding to drill at Shardin instead of Maidan-i-Naftun, which Reynolds favored. Later Maidan-i-Naftun was called Masjid-i-Sulaiman (Mosque of Solomon) after some nearby ruins that dated back to the Parthian Empire, which controlled Persia in 160 B.C.

Reynolds was now past fifty-five years old. He had been drilling in Persia for almost four years. The climate and troubles had done nothing to improve his sour nature. His letters to the home office in London became increasingly caustic, and his treatment of home-office visitors was scarcely civil. It was said that he had lost all regard for anything except punching a hole in the earth.

Reynolds's work in southwest Persia was practically a repetition of his difficulties in the northwest. Equipment could be brought up from the Gulf much easier, but the local Bakhtiari tribal chiefs, like those at Chiah Surkh, wanted their share. The Burmah Oil Company, with its ties to the British government, was not inclined to pay off as D'Arcy had done at the other location. They arranged for a British gunboat to sail up the Karun River to put the tribesmen in their place. The boat ran aground in the shallow river, and Reynolds had to make a "gold peace" with the triumphant tribesmen after all.

Work began at Shardin in December 1905, three years and one month after Reynolds began operation at Chiah Surkh. Having received permission to drill at Masjid-i-Sulaiman too, he started building a road to the other site. In the meantime his drill bits were hammering away at the Shardin rock.

The concessions syndicate received a hard blow in January 1906. The British elections brought the Liberal party to power. The party had pledged a reduction in military expenditures. Sir John Fisher was retired as First Lord of the Admiralty, and his Oil Committee was disbanded. There would be no such foolish waste of public money as converting ships to oil-burners. This action was a blow to D'Arcy, for it cut the possible market for Persian oil—if he ever found any.

Reynolds continued to fight heat, drought, insects, and incompetent labor as he pounded a hold in the ground at Shardin all through 1906. The road he was building to Masjid-i-Sulaiman ran into difficulties. It was not completed until the end of the year, delaying drilling in the Plain of Oil (so called because of the numerous oil seepages there).

The next year began with renewed trouble with the Bakhtiari tribesmen. Reynolds complained to Burmah, and the company arranged for the British government to send Lieutenant Arnold Wilson and a company of the 18th Bengal Lancers from India to keep order. Although the Admiralty had rejected the idea of converting the fleet to oil, the government—faced with the increasing popularity of the automobile—wanted a reliable British source of oil. Thus there was no argument when Burmah asked for military support for Reynolds in Persia.

Arnold Wilson later wrote a book about his experiences as a "political officer" in Persia. He spoke highly of Reynolds, telling how the aging driller defied threats of violence and was "completely single-minded in his determination to find oil." Wilson mentioned the sarcasm in the letters Reynolds wrote to the directors in London and the bitterness the driller felt toward them.

"I am not surprised at his attitude," Wilson wrote, "for his directors were unsympathetic and seem, from the letters I have read, to be not at all appreciative of what Reynolds in doing for them.''

The Burmah Oil Company's geologists had recommended drilling to a depth of 1,600 feet. The two Shardin holes reached a depth of 2,000 feet in September 1907, with no indication of oil sand. There was a conference between the geologists and the directors in London and Glasgow, Scotland, where Burmah had its headquarters. Reynolds was ordered to abandon drilling at Shardin and to move to Masjid-i-Sulaiman.

The Masjid-i-Sulaiman field was ready for drilling solely because of Reynolds' extraordinary feat of building a road under the most difficult conditions. He received no credit for this. Indeed, the directors seemed to feel that it was the driller's fault that no oil had been found. The canny Scotsmen who controlled Burmah were becoming uneasy at the way their money was being spent with no sign of any future return. One board meeting followed another while the frugal directors tried to decide what to do.

Their decision reached Reynolds in the form of a coded cable. ''Funds for drilling exhausted. Cease drilling. Dismiss staff. Dismantle equipment and transfer to coast all material worth saving.''

Reynolds was stunned. He had fought the desert, the tribesmen, and his board of directors to keep the rigs drilling. Except for these final two wells at Masjid-i-Sulaiman, he had had no say in the drilling locations. It was he who had urged this final location. The stream below the ledge where the Number 1 Well was spudded in was tinged with oil, the result of a seepage. Operating on faith and hunches, Reynolds was positive that he would strike oil this time.

Lieutenant Wilson asked what Reynolds intended to do. The old driller replied that he would wait for the follow-up letter. Business communications were dispatched by coded cable from London. Since there was always the possibility of the code's becoming garbled in transmission or in decoding, each message was followed up by another sent by boat mail.

This naturally would take several weeks to arrive. In the meantime, Reynolds told Wilson, "We'll keep drilling until the cable is verified." To his crew, Reynolds passed word that they had just about thirty days in which to strike oil or be fired.

The message to shut down was received on May 11, 1908. Two weeks later, May 26, at 4 o'clock in the morning, the earth began to rumble under the Number 1 Well. Reynolds gave a hasty order to pull the drill bit from the hole. Within minutes oil began shooting up through the casing. Reynolds yelled for the crew to run as oil shot up over the top of the derrick and fell like black rain on the fleeing crew. They had brought in a "gusher."

While the drilling crew struggled to "cap the well" and bring the flow of oil to a stop, Reynolds composed a hasty cable to the home office of the Burmah Oil Company. He informed them that he had struck what appeared to be a successful well at a depth of 1,180 feet. Details on the quality of the oil would follow in a later message. The message was dispatched by camel messenger to Basra in Iraq (then known as Mesopotamia), where it could be cabled to England and Scotland.

Lieutenant Wilson had orders to report any progress to Major Percy Cox, the British political agent in the Persian Gulf area. Wilson had not arranged a code with Cox, and he hesitated to send an open message, for he did not want to reveal the strike to any prying Persian eye. He decided that, being Muslims, they would not be familiar with the Christian Bible. So his message said, "See Psalm 104 Verse 15."

Cox read the verse, which said, "And wine that maketh glad the heart of man, and oil to make his face to shine. . . ." He immediately cabled the British embassy in Tehran, where word was cabled to the British Foreign Office in London. Reynolds' camel messenger was still plodding toward Basra when the government informed Burmah Oil Company officials that Reynolds had brought in a well.

There was guarded enthusiasm in London and Glasgow. D'Arcy summed up all their feelings when he said, "This is wonderful news—*if* it is true. I remember quite well what happened at Chiah Surkh. We had flowing oil for two weeks and then the pool ran out.''

A short time later Reynolds brought in the Number 2 Well with equally good prospects. Burmah was sufficiently encouraged to authorize Reynolds to move the Number 1 rig and try a third hole. There were no tank farms or pipelines to handle the production, and the successful wells had to be capped. However, the company wanted to define the expanse of the field. This would give their geologists some idea of the size of the underground oil pool. From this they could make a fairly good estimate of the amount of oil that could be recovered.

The so-called oil pool—which is rarely a pool at all—has no regular shape. It is simply an underground area where oil is trapped, often soaked in porous rock like water in a sponge. Its expanse and depth depend upon the position of the nonporous rock that imprisons it.

The three successful ''offset'' wells, or wells opposite each other, proved that the company could afford to risk going ahead with development on a large scale. However, striking oil was only the beginning. The major problem now was what to do with the oil they had. This meant constructing pipelines to the coast—a job that proved almost as formidable as finding oil in the first place. Next, the company had to decide if it should ship the crude oil to England or Europe, or if it should construct a refinery in Persia. They decided to build a refinery at Abadan, a low island where the combined Tigris and Euphrates rivers flowed into the Persian Gulf. This was a fateful decision the company was to regret years later when rampant nationalism threatened to drive British oilmen from the fields they had worked so hard to develop.

3
Winston Churchill
Takes a Hand

Reynolds' discovery of oil did not suddenly make everyone concerned rich. People knowledgeable in the oil business estimated that it would be five more years at best before Burmah Oil Company would see a single shilling returned on its investment in the Persian drilling project. Worse yet, the company would have to spend many times what it had already sunk into the wells to build pipelines, tank farms, and a refinery. Persian oil was useless until it could be brought to market.

However, world demand for petroleum for lighting and for automobile fuel was increasing dramatically each year. The directors of Burmah Oil could foresee enormous future profits from their Middle East oil venture. They readily subscribed the necessary capital to develop the field. In the spring of 1909 they formed the Anglo-Persian Oil Company to handle the new field. Burmah Oil took all the common shares of stock, valued at $4 million. D'Arcy received about $1 million and Burmah Oil stock worth $1 million more, which repaid him for his initial capital losses. He was also made a director of Burmah Oil. The company then sold to the public 600,000 shares of preferred stock, guaranteed to pay

6 percent interest for five years, in order to raise capital to begin the necessary pipeline and refinery. These were bitter days for C. B. Reynolds. He had done a superhuman job in finding oil in the barren Persian hills. Now he wasn't needed or wanted anymore. During the first difficult years he had drilled in Persia, Reynolds had run his own show, because none of his bosses in London knew enough about the operations to give him many orders. Reynolds naturally expected to continue to boss the operations. Instead, he found himself set aside by experts and managers sent out from the head office in London. No more drilling was authorized until the refinery was ready, which left Reynolds with practically nothing to do. He sat in his office, drank his liquor, and cursed his bosses and those they had sent out to bother him.

Then the British government began meddling in affairs. The Foreign Office in London decided that negotiating with the Persians was too important a matter to be left to the company. Major Percy Cox, the British political resident for the Persian Gulf area, was instructed to negotiate with the local sheikh for land for the· refinery. The Shah in Tehran was ignored.

On May 6, 1909, Cox signed an agreement with the sheikh, which transferred one square mile of Abadan Island to the Anglo-Persian Oil Company in exchange for a 6 percent royalty on oil refined there. This royalty was to be deducted from the 16 percent royalty due the Persian government for oil produced from the fields. The Persian government was not consulted in this high-handed giveaway of its royalties, an act, both illegal and immoral, that set the tone for future relations between the Anglo-Persian Oil Company and the Persian government. The seeds were sown right then for the future discord that would eventually drive the British out of Persia. It also caused nervousness among other Middle East nations, which were worried about British imperialism.

Abadan Island sits at the end of the Persian Gulf where the

Tigris and Euphrates rivers converge to form the Shatt al-Arab River. The island is almost pure sand, built up as part of the river delta. When Anglo-Persian moved in, it was unpopulated except for date farmers and some fishermen in villages along the shore.

A dock was hastily constructed on Abadan, and ships began unloading cargoes of pipe. Reynolds thought he would oversee construction of the pipeline. The company picked Charles Ritchie instead. Ritchie, a Scotsman as hard-bitten as Reynolds, had already had considerable experience in pipelining in Burma.

Ritchie was hired only to lay out the line route. He took an assistant, J. A. Jameson, two native porters, and two mule loads of stakes. They struggled up rocky defiles, crossed barren areas, slipped along steep ridges, and struggled through deep streams. At intervals Ritchie stopped to drive stakes in the ground. Strips of red calico were tied to the stakes so they could be easily followed.

There were no civil engineers with transit compasses. Ritchie worked everything out with his eyes and his head. Jameson said later that he could not at the time see how any man could lay pipe over the wild route Ritchie marked out. Other pipeliners agreed. The company, in desperation, talked Ritchie into staying on to lay the pipe himself, although he did not want the job. He asked Jameson to stay with him. Jameson agreed to stay one year but remained fifteen.

Ritchie had all of Reynolds' troubles except difficulty with the natives. The local sheikh, satisfied with his royalty payments, kept his people under strict control. This permitted the British government to withdraw the Indian army detachment under Lieutenant Wilson. Wilson later resigned from the army and returned to work for Anglo-Persian.

The pipe was brought to Abadan by freighter. There it was loaded on a creaking river steamboat and brought within seventy-five miles of Fields (as the oil field was known). From this point, tough little Persian mules, almost as small as donkeys, dragged the pipe to its position along the pipeline route. Roustabout crews of natives, Persians and Arabs, rolled each piece of the pipe until it butted against the previous joint. An iron collar was screwed on to join the two pieces, and then the pipeline crew used huge pipe tongs to tighten the joint into the collar. The foreman timed their actions by shouting, "Ho! Ho!"

Legend claims that Ritchie knew only one local word, an Arabic one, *Yella!* which means "Get moving!" With this single command, he strung 150 miles of iron pipe across the barren ridges from Fields to Abadan. Where possible, the pipeline was buried in the ground. On the rocky uplands and ridges, where ditching was impossible, Ritchie laid the pipe on the ground and kept going.

As the pipeline inched slowly along, work began on the refinery at Abadan. Geologists laid out future drilling sites. Sheikh Mohammed, who had signed the agreement with Major Cox, became increasingly worried at the anger of the Qajar government in Tehran. Cox assured him that the British would not let the Tehran government remove him as local ruler.

In the meantime, Reynolds sat with nothing to do. Managers sent out from London made all arrangements. Finally, in early 1911, a management survey from Anglo-Persian headquarters recommended to the home office that Reynolds' contract be settled "on the best possible terms." The survey acknowledged his extraordinary achievement in drilling for oil under the most trying conditions, but reported that neither his attitude nor his ability fitted him to boss the continuing development of the Persian oil fields. An agreement was reached, and Reynolds left the company's employment in February 1911. He never revealed the terms of the settlement, but it must have been substantial. He returned to England, where he sported a monocle, dressed in the latest fashion, and was often seen at the racetracks. Those who remembered his Spartan life in Persia were dumbfounded when they chanced to meet him in England.

Ritchie finished his pipeline, laying two lines, one of six-inch and the other of eight-inch pipe, by the middle of 1911. The line, like Reynolds' wells, had to wait for completion of the refinery on Abadan Island. Meanwhile, the sand and mud flats of the island were being slowly transformed. Houses for the engineers and technicians had been built. Barracks and shacks were constructed for the Arab and Persian work crews. Huge tanks and cylinders of steel rose above a spider web of steel braces in the refinery area. Eventually the Abadan refinery became the largest in the world, and remains so today.

Petroleum is a combination of hydrogen and carbon, with various impurities. When broken into its various parts (fractionated), petroleum produces a variety of products, such as gasoline, fuel oil, kerosene, lubricating oil, asphalt, and petroleum jelly (Vaseline), among others. All of these individual items are called fractions of petroleum. The refining process breaks the petroleum into these fractions.

In its basic form, a refinery is a distilling plant. The crude oil, brought by pipeline or tanker from the fields, is heated to a vapor. Each of the fractions vaporizes at a different temperature, so that the refinery can "boil off" each fraction separately.

The vapors pass from the heater section into a huge upright fractionating tower (bubble tower). The vapors rise in the tower and condense back to a liquid at different levels. The liquid is then drawn off by pipes.

Gasoline is the first fraction to vaporize. It is lighter than kerosene, fuel oil, and lubricating oils. It rises higher in the fractionating tower and is drawn off at the top. Other fractions are drawn off at lower levels. Heavy residues at the bottom become tar, asphalt, paraffin waxes, and petroleum jelly. Nothing is wasted. There is a use for each petroleum fraction.

This refining process is known as "thermal cracking." After 1936, "cat cracking" (catalytic fractionating) came into use. This process uses a catalyst to help break down the fractions. Cat cracking produces a higher grade of gasoline than straight thermal cracking.

Building a refinery is a slow process. Abadan was not ready for

Abadan on the Persian Gulf became the world's largest oil refinery.

operation until late 1912 and did not get into full production until early in 1913. Eleven years had passed since D'Arcy first sought a drilling concession in Persia, and almost five years since Reynolds brought in the discovery well to prove the tremendous possibilities of Middle East oil fields.

World politics were beginning to affect the future of Middle East oil drastically. The British admirals continued to drag their feet on conversion of the entire fleet to oil, but they were forced into action when Winston Churchill became First Lord of the Admiralty in 1911.

There was no longer any question about oil's superiority to coal as a ship's fuel. In fact, the British navy had built fifty-six oil-burning destroyers and seventy-six oil-fueled submarines. However, the backbone of the fleet was its giant battleships. These could not be converted until Britain had an ensured source of petroleum that could not be cut off in wartime or overpriced by a monopoly in peacetime.

Churchill appointed a Royal Commission on Oil Supplies. Admiral Fisher was recalled from retirement to head the commission. Fisher immediately sent a four-man subcommittee to Persia to determine if Anglo-Persian could supply the total needs of the British navy if complete conversion to oil was carried out.

The group inspected the Abadan refinery and heard reports on its present and future capability. Then they went to Fields, where oil geologists showed them how new wells lately drilled proved that the oil pool was larger than any of them had dreamed it would be. They explained how they estimated proven reserves from the information gained in studying these offset wells. They insisted that their estimates were conservative. Even if no more oil pools were discovered, the proven oil reserves under Anglo-Persian control could supply the total needs of the British navy for at least fifty years.

The subcommittee's report convinced the commission. However, Lord Fisher was outraged when Winston Churchill, in an address to Parliament on July 17, 1914, suggested that the Admiralty become "the independent owner and producer of its own supplies of liquid fuel." This meant simply that Churchill wanted the government to take over Anglo-Persian. Fisher, who believed that government ownership was socialism, bitterly denounced Churchill.

In the oil dispute, Churchill displayed the same determination he later showed as Britain's great World War II prime minister. He went ahead and introduced a bill in the House of Commons for the government to buy a 51 percent interest in the Anglo-Persian Oil Company.

Surprisingly, the company's directors did not oppose the stock sale to the government. Churchill had privately assured them that the purchase was solely to guarantee that the company would remain under British control. He also promised that the government would not interfere in the actual operations of the company. Business decisions would continue to be handled by Anglo-Persian's regular board of directors.

There were other considerations that were not publicly discussed. One was that government ownership of the oil company would ensure that the British navy and army would be available to counter any attempt of the Persian government to cancel the concession or to pressure the company for increased royalties. Already there were rumblings of dissatisfaction in the Qajar government over some of Anglo-Persian's actions in the oil-field area. For one thing, company employees treated the native Persians as if they were second-class citizens.

A secret agreement was drawn into the bargain. This agreement stated that the company would sell oil to the Royal Navy at a price well below the world price, and it had the effect of defrauding the Persian government of royalties, because at that time royalties were figured upon a basis of company profit rather than barrels of oil produced, as was done later. The secret price reduction to the Royal Navy lowered the Persian revenue from the wells.

Agreements such as this, plus reducing the Persian government's share of the Abadan royalty by the amount paid to the local sheikh, systematically deprived the government of a share in its own natural resources. Within a decade this shortsighted policy began a chain of events that eventually opened the way to the Iranian-Arab oil monopoly, which today has a stranglehold on world oil prices.

Persia lost a considerable amount of money in the secret navy oil-purchase agreement. Winston Churchill, writing in 1923, said that through this contract the British government saved about $35 million, or about half the money it had put into Anglo-Persian to buy control in the first place.

While the oil company welcomed the government as a partner for political reasons, it also welcomed the $70 million in cash that the stock sale brought into its treasury. Since its organization in the spring of 1909, Anglo-Persian had operated at a complete loss. The company had poured millions into the pipelines, the refinery, and the new drillings, while unable to sell any oil until the refinery was completed. Though Anglo-Persian expected eventually to make a handsome profit on the enterprise, it was badly in need of cash. In fact, one of Churchill's arguments to Parliament was that the company faced bankruptcy. If this should happen, he pointed out, there was a possibility that the Rothschilds in France or the Rockefellers in the United States might buy in and gain control, just as Great Britain had in 1875 become the largest shareholder in the Suez Canal Company by buying the shares of the Khedive of Egypt.

Churchill signed a tentative argreement to buy the stock in May 1914. The House of Commons, dominated by a Liberal government, ratified the agreement on July 7, but failed to vote the necessary funds. Then World War I broke out on August 4. The House hastily voted the money the next day.

With a major world war in progress, the British government had a difficult job facing it in the confused political mess in the Middle East. Occupation troops were moved into southern Persia to protect the oil fields. Russia, fearful that Great Britain intended to take over all of Persia, occupied the northern provinces. The weak, corrupt Qajar government of Persia was powerless to do anything about the foreign occupation.

The rest of the Middle East was not so easily handled. The Turkish Ottoman Empire, centered in Constantinople (present-day Istanbul), Turkey, controlled Syria, Palestine, and Mesopotamia, along with parts of the Arabian Peninsula. Turkey joined Germany in the war against England and France. That put the enemy at the borders of Persia, distinctly threatening the oil supply upon which Great Britain now depended.

The Ottoman Turks had begun their rise to power in the Middle East by destroying the Byzantine Roman Empire in 1453 and occupying the capital, Constantinople. The Ottoman Empire then spread into Bulgaria, Greece, Serbia (now Yugoslavia), Albania, Romania, and Hungary in Europe. In Asia Minor the Ottomans captured what are now Syria, Lebanon, Iraq, Jordan, and Israel. They then captured Egypt and all of North Africa to Morocco, plus most of the Arabian Peninsula coast.

By 1914, when the Turks joined Germany in the war against France, Italy, and England, the Ottoman Empire had long since declined. The Turks had been expelled from Europe, North Africa, and Egypt. However, they still held the land from Turkey to the borders of Persia and Egypt, along with sections of the Arabian Peninsula coast.

The bulk of British troops was bogged down in trench warfare in Europe. Large numbers of soliders could not be spared for the Middle East. The situation became especially grave after the British defeat by the Turks in the Gallipoli Peninsula campaign of 1915. Churchill was blamed for pushing this battle and was forced to resign from the government.

The next year, 1916, the British suffered another blow when 8,000 troops surrendered to the Turks in Syria. This second defeat set off a chain of political actions that had a heavy impact on the future of oil in Mesopotamia. The situation was so grim that the British sought French aid. A month after the surrender the British government promised to give France a section of the Ottoman

Middle East empire as part of the final war settlement. This section included present-day Syria, Lebanon, and the northern area of Iraq.

The British expected the French to put troops into the Middle East to help fight for its new empire. The French accepted the gift, but supplied no troops.

The bulk of Middle East fighting continued under the British, aided by the Arabs organized by Sheikh Faisal and the famous Lawrence of Arabia. T. E. Lawrence was an archaeologist who helped organize Arab guerrilla teams to fight the Turks. The Arabs fought with the understanding, encouraged by Lawrence, that they would have their independence at the end of the war.

The Arabs were betrayed. The victors divided the Middle East among themselves when the war ended. The newly formed League of Nations nominally took possession of the former Ottoman Empire, but delivered it to the French and British to govern under "mandates." The mandate was supposed to last until the mandated country was ready for self-government, which in effect meant as long as the occupying government chose to stay.

The British now realized that they had made a mistake in giving so much of the Middle East to France in 1916. In the first place, Britain got nothing for the gift. Secondly, it now appeared that the area given to France in northern Iraq had prospects of oil deposits as rich as those in Iran. This was a serious loss, because World War I had shown the enormous value of oil in modern warfare. Lord Curzon, the British statesman, said, "The Allies floated to victory on a wave of oil."

The British began a campaign to regain the lost territory, causing a conflict between them and France that worked to Turkey's advantage. The Turks were now able to play the two rivals against each other.

This shows how international politics and the oil business have always been as closely intertwined as threads in a piece of cloth. In using oil as a political weapon today, the Arab oil-producing nations are only carrying on a long tradition started by the British, French, and Germans years ago.

4

The Land
Between the Rivers

Present-day Iraq stretches from the Turkish border south to the Persian Gulf. On the east it butts against Iran (Persia), and in the west it touches Syria, Jordan, and Saudi Arabia. It also adjoins Saudi Arabia and Kuwait in the south. Cutting through Iraq are the fabled Tigris and Euphrates rivers, which join together to make the Shatt al-Arab River that flows into the Persian Gulf at Abadan.

The area between the Tigris and Euphrates was one of the cradles of civilizations. The Greeks called it Mesopotamia ("the land between the rivers"). It continued to be known by this name until after World War I. Settlers came into Mesopotamia before 4000 B.C. The original inhabitants were followed by the invading Sumerians, who build a great civilization between 4000 and 2700 B.C. The Sumerians gave way to the Semites, who developed a series of great empires, including the Akkadian, the Babylonian, the Assyrian, and the Chaldean.

This was the land of the early Bible. Biblical historians place the Garden of Eden here. Ur of the Chaldees, the ancestral home of Abraham, patriarch of the Children of Israel, was on the

Persian Gulf. This was the home of Noah, for the Gilgamesh records of Nineveh told the Noah story long before it was taken into the Holy Bible. Mesopotamia was also the land of the Daniel and other biblical heroes.

In 539 B.C. the Babylonian Empire fell to Cyrus the Great of Persia, and it was then conquered by Alexander the Great in 334 B.C. In the passing years Mesopotamia was ruled by the Romans, Sassanians, Arabs, Mongols, and then the Ottoman Turks, who came to power in 1638. Mesopotamia rose and fell in international importance during those years. But when World War I ended, Mesopotamia's great days were long behind it. Still, the promise of great oil deposits gave the "land between the rivers" a new importance that oil-hungry European nations could not ignore.

The Germans were the first to take an interest in possible petroleum deposits in Mesopotamia. Before World War I they began pioneering the idea of fueling fleets with oil. Germany had no oil deposits within its own territory and was able to import only 125,000 tons annually from Hungary. The rest of Germany's oil needs before the war were supplied by Rockefeller's Standard Oil Company. These imports from the United States satisfied about 91 percent of Germany's oil needs.

The German high command realized that Germany's imperialistic plans would lead to war with Great Britain and France. It was equally clear to them that oil would play a very important part in fueling any modern war machine. Germany would be greatly handicapped if it had to depend upon American oil sources in a war with Great Britain. In the event war did not come, which was unlikely in German high-command thinking, Germany needed an oil source of its own to escape high American prices.

This situation caused the German government to take a great interest in developing oil fields in Mesopotamia, just as the British were doing in Persia through the politically dominated Anglo-Persian Oil Company.

Up to this time no indications of oil had been found in any Middle East country except Persia and Iraq. The British, through

D'Arcy's pioneering, were solidly entrenched in Persia. So the Germans turned to Iraq. Clues to hidden oil were the same there as in Persia: numerous seepages and tar pits. In addition, in the Kirkuk area, south of Mosul and the ruins of biblical Nineveh, there were the "eternal fires." These fires, known to have been burning since the days of Babylon, were fed by natural gas seeping from the earth. The gas had probably been ignited by lightning in prehistoric times.

The Turks were also interested in developing the oil potential in northeastern Mesopotamia, which they still controlled at that time. However, Abdul-Hamid, the Turkish sultan, lacked both money and experience, so he transferred all the possible oil land in the empire to his private control and tried to interest foreign companies in developing fields.

In the years that followed, various companies tried to work out oil concessions in Turkish territory. None succeeded. Then in 1908, the year Reynolds brought in the first big Persian well, Abdul-Hamid was driven from the throne by a revolt of "Young Turks." The oil companies began negotiating all over again with the new sultan.

The Germans, however, were ahead of them all. Previously a German company had begun building a railroad from Turkey to Baghdad. The company, as an inducement to finance the road, was given rights to prospect for minerals in areas along the route. The company's engineers, guided by secret geologists' reports, deliberately planned the railroad so it would include the Kirkuk region.

Another company seeking oil concessions in Iraq was the Royal Dutch–Shell Company. The new firm was a combination of the Royal Dutch Company, which owned oil concessions in Indonesia, and the Shell Transport and Trading Company, a British firm that traded in seashells. The familiar Shell gasoline sign seen in the United States had its origin in the Shell Transport and Trading Company's seashells.

Oil exploration in Mesopotamia stopped at the outbreak of

World War I. When the war ended in November 1918, British troops occupied Syria and northern Iraq, areas that had been given to France in 1916. The British government suggested that France should give the land back, since the French had taken no part in fighting for the area.

Georges Clemenceau, the fiery French premier, indignantly brushed aside the suggestion. He knew of the oil possibilities in Mesopotamia and particularly in the Kirkuk region. France had no oil fields of its own, and Clemenceau knew from first hand experience how vital oil was to a fighting nation. In December 1917 he had written to President Woodrow Wilson of the United States:

> At the decisive moment of the war, when the year 1918 will see military operations of the first importance begun on the French front, the French army must not be exposed for a single moment to a scarcity of the petrol [oil] necessary for its motor lorries [trucks], aeroplanes, and the transport of its artillery.
>
> A failure in the supply of petrol would cause the immediate paralysis of our armies, and might compel us to a peace unfavorable to the allies.

The French message went on to complain that American oil companies were sailing tankers in "the Pacific instead of the Atlantic Ocean."

> President Clemenceau personally requests President Wilson to give the necessary Government authority *for the immediate dispatch of these tank steamers to French ports.* If the Allies do not wish to lose the war, then, at the moment of the great German offensive, they must not let France lack the petrol which is as necessary as blood in the battles of tomorrow.

And since he felt that oil was as "necessary as blood," the French leader flatly refused to return the prospective oil lands to the British.

David Lloyd George, who had become prime minister, re-
taliated by ordering all British occupation troops out of Syria after
first establishing Amir Faisal as the head of the government.
Faisal had been commander-in-chief of the various Arab guerrilla
bands brought together by Lawrence of Arabia.

The French were horrified at the threat of British removal from
Syria, for France was in no position to move troops into the area.
Lloyd George blandly replied that no French troops would be
necessary. Their very good friend Amir Faisal would keep control
without trouble. The French did not trust the amir, and for very
good reasons. In March 1920, Faisal convened a "Syrian National Assembly," which immediately elected him king of Syria.

To make matters worse, the Turks—who had kept their army
intact after the surrender—now disavowed the armistice and invaded Syria. The French, despite the poor condition of their army
after four bitter years of fighting the Germans, were forced to fight
again. They opened with a successful attack on Damascus, Faisal's capital. He fled to the British for protection. This supported
the French claims that Britain was behind French problems in
Syria.

The trouble with Turkey was not so easily settled. Turkey, as
Germany's ally, had surrendered when Germany collapsed, but
the Turkish army was far from defeated. All through 1920 and
into 1921 they successfully fought off French attempts to drive
them out of Syria, and for a supposedly defeated people, the
Turks put up quite a stubborn battle.

The war with Turkey was extremely unpopular in France. The
French people, weary of war, saw no reason for continued fighting, and after a year, pressure forced the government to make
peace with the Turks.

There was considerable bitterness in French military and government circles. They blamed the British for starting the Middle
East trouble, claiming it was done to recover the Mesopotamian
oil area. It appeared that France would have to abandon the
Mosul-Kirkuk area, but they were determined that the British

would not profit at France's expense. So General Gouraud, chief
negotiator for the French, met with Turkish delegates in February
1921 to discuss peace terms. He surprised the Turks by offering to
give them back the area containing the prospective oil fields. He
also agreed to withdraw French occupation troops from southern
Turkey. The Turks readily agreed, and the war was ended in
March, 1921.

The British were outraged. The terms of the armistice called
for Allied occupation of portions of the former Turkish empire.
France's action in returning portions to the Turks was a clear
violation of the armistice agreement, Britain argued. They
claimed that France had no right to make any concessions to the
Turks. The French and British disagreements deepened, and the
friendship of World War I disappeared as self-interest took over.

Under the Treaty of Sèvres at the end of the war, Greece was
given a mandate to govern the Turkish province of Smyrna as its
part of the Ottoman Empire spoils. Turkey refused to sign the
treaty or to permit the Greeks to occupy Smyrna. This reopened
the war between Greece and Turkey. Great Britain immediately
backed Greece with arms and money. The French, bitter at the
British, gave similar aid to Turkey.

Anton Mohr of the University of Oslo, Norway, made a study
of oil and politics during this period. After Turkey badly defeated
the Greeks, Mohr wrote, ''For the British government, the defeat
was almost as stunning a blow as it was for the Greeks themselves. The whole of the oriental policy of Lloyd George was
based on the assumption that the Greeks would win.''

Lloyd George tried to rally the British people for war with
Turkey. He claimed that ''this involves nothing less than the loss
of the whole results of the victory over Turkey in the late war.''
The British people were as weary of war as the French. They
rejected the prime minister's appeal, and his government collapsed
in October 1922.

The new British Prime Minister, Bonar Law, was equally
determined to secure the Mesopotamian oil land. While the Kir-

kuk field had not been proven, exploratory wells sunk in 1919 indicated good prospects.

Frustrated in its desire to take military action against the Turks, Great Britain swung a political and diplomatic club. It first formed the nation of Iraq from its mandated Mesopotamian area and made Amir Faisal, the erstwhile ''king of Syria,'' king of the new state. The northern border of Iraq was left open because of the unsettled political situation.

According to Mohr, ''The British hoped that the boundary would be drawn so far north as to regain possession of all the districts 'maliciously' conceded by the French to the Turks in March 1921. This hope was now disappointed, and the Turks were more powerful and difficult to negotiate with than before.''

The British position was made worse by the appearance of Admiral Colby Chester, who represented a group of American companies. Chester had previously served in Turkey in 1898 as a member of the American embassy in Constantinople. He had returned in 1908 to try to obtain oil leases. Now he was back in another attempt to secure the Kirkuk leases.

At this point the British called for a conference in Lausanne, Switzerland, to arrange a final peace treaty, which they hoped would head off American ambitions in Mesopotamia. Lord Curzon, the British negotiator, gave the Turks an ultimatum that included an order to withdraw from northern Mesopotamia. At the same time, the British made the mistake of objecting to France's occupation of the German Ruhr industrial district. This irritated the French to the point where they supported the Turks against Curzon's ultimatum. The Turks then refused to sign the treaty.

The conference broke up, but was recalled in 1923. This time a peace treaty was signed, but it proved a victory for the Turks. They regained control of the Dardanelles and Bosporus straits, and foreign troops were removed from Turkish lands. On the disputed Mesopotamian border, the treaty said only that Turkey and Great Britain would settle the border by negotiation. If this could not be done within nine months, the question would be

"Blowing wild"—the discovery well in the Kirkuk fields, Iraq, spewing oil above the top of the derrick in 1927.

Courtesy Iraq Petroleum Company

referred to the League of Nations for settlement. Turkey was forced into making this concession because both France and the United States had withdrawn the support they had been giving Turkey.

Turkey and Great Britain, as everyone expected, could not agree. The dispute was referred to the League of Nations, which appointed a commission to inquire of the people whether they wanted to be ruled by Turkey or Iraq. In December 1925, the commission awarded the disputed land to Iraq, with the provision that the League's mandate to Great Britain to control Iraq would continue for twenty-five more years. Turkey received $25 million as compensation for the oil it thus lost.

At the same time, the Turkish Petroleum Company (controlled by Anglo-Persian, Shell, and French interests) was allowed to resume its old concessions to drill in the Mosul area. However, the stock was increased to give a consortium of seven American companies a quarter interest. Shell, Anglo-Persian, and a French consortium each held a quarter interest. Under these circumstances, the Turks had good reason to believe that France and the United States had sold them out for an increased share in the oil.

Earlier than this, in 1923, a well drilled at Naftkhana in the Mosul area had proved the definite existence of oil in commercial quantities. However, the big strike came in 1927, after a series of disappointing holes. The Number 1 Well drilled at Baba Gurgur, north of Kirkuk in northeastern Iraq, came in "blowing wild" on October 15, 1927. The "gusher" came in with a violence that spurted oil forty feet over the top of the derrick. Crews fought the flow for a week before they got a cap on the casing and brought the well under control.

The discovery indicated an oil strike of enormous importance. Drilling was discontinued in other sections of Iraq in order to put down additional wells in the Kirkuk area to determine the expanse of the new pool. These exploratory wells showed that the oil was trapped in an anticline. That is, the oil was in limestone rock that

Sections of pipe being fastened together and dragged across the Euphrates River in Iraq, as part of the pipeline from the Iraqi oil fields to the Mediterranean coast.

Courtesy Bechtel Corporation

peaked in the center and sloped down on each end. The oil-bearing rock was narrow, but extended almost sixty miles through three distinct domes. The domes were covered by earth folds created during mountain building.

As when Reynolds brought in his discovery well in Persia, drillers found oil in plenty, but had no place to send it.

The Turkish Oil Company (which wasn't Turkish at all) went through several name changes and finally became the present Iraq Petroleum Company (IPC). After the Kirkuk field was discovered, the company built a small refinery to supply Iraq's own needs. Then construction began on two 12-inch pipelines to the Mediterranean seacoast. One line, 532 miles long, went through Syria and Lebanon to Tripoli. The other, 620 miles long, ran from Kirkuk to Palestine. The first line was opened in January 1935.

The Palestine line became a casualty of the Arab-Israeli conflict after World War II when the Iraqi government closed the valves to prevent the oil from going into Isreal. The line is still closed.

In the meantime, trouble has developed in Persia that foreshadowed the almost disastrous oil conflicts of the 1950's.

5

Trouble on Oily Waters

Oil will calm troubled waters. Tankers, threatened by heavy seas, have sometimes poured part of their cargo into the ocean to reduce the wild waves. To pour oil on troubled waters has become a cliché for smoothing over a difficult situation.

The sea, however, is in the physical world. In the political world, oil is not a calming influence. It is itself the cause of troubled waters.

In Persia the situation seemed well in hand. The Anglo-Persian Oil Company had a sixty-year concession. The British government held 51 percent of the company's stock, which would appear to stall any attempt of the Persian government to interfere in the company's operations. Great Britain was one of the world's three most powerful nations and floated the world's largest navy. It had always been a matter of national policy for the navy and army to support British economic ventures throughout the world. Under these conditions, the directors of Anglo-Persian felt quite secure.

Before many years passed, this security was shaken as the Persians began to demand more profit from their oil. The battle between the oil owners and the oil exploiters moved beyond Persia into the realm of international politics.

The titanic struggle between Persia and the Anglo-Persian Oil Company was closely watched by other Middle East countries, which learned that they could successfully demand greater and greater shares in their own oil. This led directly to the present oil cartel controlled by the oil-producing countries of the Middle East—a situation that quadrupled world oil prices in 1973.

Trouble between the Anglo-Persian Oil Company and Persian officials began in 1915. Up to the beginning of World War I the company had poured millions of dollars into building up the oil industry in Persia without receiving a cent in return, but the outbreak of the war in August 1914 turned Persian petroleum into black gold. Profits started flowing into the Anglo-Persian treasury. At the same time, the company began a policy to reduce Persia's already inadequate share in the oil revenues.

The Qajar ruler was powerless to do anything about the situation, especially after Russia and Great Britain occupied portions of Persia. This happened after Turkey entered the war on the German side. The Czarist government of Russia moved into northern Persia to counter the possibility that Turkey might use Persia as a corridor to attack Russia's southern oil fields in the Caspian Sea region. Great Britain, claiming it was similarly threatened, occupied the southern area of Persia to protect the Anglo-Persian concessions.

The only real threat to Anglo-Persian operations came in 1915 when local tribesmen, bribed by German and Turkish secret agents, cut the pipeline between Fields and the Abadan refinery. All refining operations came to a halt for five months, causing a loss of 144,000 tons of gasoline.

The company had made a previous arrangement with the local sheikh for protection, paying him illegally from the Persian government's share of oil royalties. However, instead of blaming him for the attack on the pipeline, company officials blamed the central Qajar government. They announced that all losses would be charged to the government's royalty. These losses amounted to over $2 million in damages to the pipeline, plus an additional million due to lost production while the refinery was shut down.

The Persian government was infuriated. They could do nothing, however, except ask that the dispute be arbitrated. The company refused, although arbitration was provided for in the D'Arcy concession agreement.

The Russian forces withdrew from Persia after the Communist Revolution of 1917. However, the Persian Cossacks—the main force of the national Persian army—had been organized and were still officered by White Russians. These officers used the Qajar troops to make attacks into Red Russian territory. The Communists claimed these attacks were inspired by the British. V. I. Lenin, the Russian revolutionary leader, then reoccupied northern Persia to stop the Persian Cossacks' raids. This reoccupation alarmed the British Foreign Office. Russia had been a war ally of Britain and France until the Red Revolution.

Lenin then arranged an armistice with the Germans, and Britain feared that Russia might use its idle Red Army to gain all of Persia. Hasty negotiations were started with Lenin. They resulted in an Anglo-Russian treaty that defined Soviet and British spheres of influence in Persia.

The Persians defined this treaty for what it was: the beginning of a division of their country between the two powers. There was increasing public unrest against the Qajar government for its failure to do something about the foreign takeover of their country.

This dissatisfaction led directly to the revolution of 1921. Sayyid Zia al-Din Tabatabai, a fiery young intellectual, was the leader. He had worked as an undercover agent for the British and had their full support in his revolt against Qajar rule. The British could not risk Soviet displeasure by taking a direct role in the revolt, but they could back Sayyid Zia al-Din secretly and assure him of the backing of the Persian Cossacks.

Russia played into British hands by joining Great Britain in pressuring the Qajar shah into dismissing the Cossacks' White Russian officers. This brought Colonel Reza Khan into command of the Cossacks. Reza Khan, who had already been brought into the revolutionary plot by Sayyid Zia al-Din and British secret agents, was supposed to bring other key Persian officers into the plot and then lead the Cossacks in a march on Tehran, the capital.

Reza Khan was a commoner whose ancestors had all been soldiers. Urged by his mother, who was fiercely proud of his family's military heritage, the orphan boy had joined the Cossacks as a private soldier when he was fourteen years old. He soon gained a reputation for being a very good soldier, but was held back by illiteracy. When he was twenty, in 1898, he realized that his inability to read and write would prevent him from rising in the army, so he hired a fellow soldier to teach him to write. After that, his rise was rapid. By 1919 he was a colonel, the highest rank a Persian could hold under the Russian commanders of the Cossacks.

General Harold Dickson, the British occupation commander in southern Persia, believed Reza Khan to be pro-British. The general could not have been more wrong. Reza Khan was pro–Reza Khan.

Reza Khan rallied the young dissatisfied officers of the Cossack brigade behind him and, on February 20, 1921, led the march on Tehran. The only opposition was from the Tehran police, and it was short-lived. The plotters were not strong enough to abolish the monarchy. The Persian peasantry had always had a king. They could be expected, in their ignorance, to rally around the monarch. The revolution was still too unstable to risk arousing any more opposition than necessary.

The Qajar shah nominally remained as head of state, but Sayyid Zia al-Din became premier and Reza Khan was made minister of war and commander in chief of the Persian armed forces. Sayyid Zia al-Din lasted four months, in office, and was then pushed out by Reza Khan, who was content for a while to rule through puppet premiers, whom he made and broke at will.

At the same time, Reza Khan began bringing the various

provinces back under central authority. Ignoring Soviet displeasure, he broke Communist attempts to control some northern provinces. Returning to Tehran after this campaign, he was hailed as a national hero. He then turned south, crushing one resisting provincial ruler after another.

Officials of the Anglo-Persian Oil Company became alarmed. Most of the British occupation troops had been withdrawn to Iraq. The company urgently requested that they return. The officials were uneasy at the way Reza Khan was crushing the independent local khans. The company had been doing business with the local rulers to keep peace. The local sheikhs were easier to deal with, especially since tough Reza Khan had taken over the central government.

A crisis developed when Reza Khan marched south to punish the Sheikh of Mohammerah. The sheikh had made an agreement with Sir Percy Cox when he ceded Abadan to the British that he would be protected from the central government. With this backing, the sheikh refused to pay taxes to the Tehran government. Arthur C. Millspaugh, the American financial expert hired by the Tehran government, recommended that the sheikh be forced to pay his taxes. Reza Khan agreed, marching south to enforce this.

Oil company officials hastily asked the British government to aid the sheikh. Reza Khan hesitated, but pushed on when he learned that the sheikh had sent the Qajar shah a telegram urging the shah to join him in a revolt against Reza Khan. The shah was in London, where he had fled the previous year out of fear of Reza Khan.

The shah declined to come back and the British Labor government also declined to send troops to the sheikh's aid. The sheikh was taken back to Tehran, where he was kept under house arrest until his death.

Reza Khan became so popular that he was able in 1925 to force the Majlis—the Persian assembly—to dethrone the absent shah. Reza Khan (chief) then became Reza Shah (king).

The new ruler, taking his cue from Mustafa Kemal (later Atatürk), who was modernizing Turkey, set out to remake Persia in a modern image. He needed vast sums of money for this, and he was determined to get it from Anglo-Persian oil. The company was just as determined that he would not.

The fight between the oil company and the government for more oil royalties had begun earlier, under the Qajar administration. The Persians had very valid objections, which Anglo-Persian brushed aside.

One major objection was the development of Anglo-Persian subsidiaries in Europe, totally financed by the oil company. Since the 16 percent royalty due Persia was based upon the *net* profit of the company, the expenses of building the subsidiary companies came out of the profit of Anglo-Persian. Expanding the company in this manner would have been fine if all concerned had shared in the total profits. However, Anglo-Persian took the position that these new companies were outside Persia and that, therefore, the Persian government had no right to a share of their profits. The Persians, who saw their own royalties reduced to finance these new companies, could not see why they did not get their 16 percent of the additional profits.

Another objection was Anglo-Persian's pricing system. The company reduced the price of crude oil to its own refineries outside Persia. This increased the profits of refineries whose receipts were outside the Persian royalty area and reduced the profits of the production companies inside Persia, very effectively increasing Anglo-Persian's take while cutting that of the Persian government.

The oil company absolutely refused to discuss Persia's objections. The Persians then agreed to the appointment of an Englishman, Sydney Armitage-Smith, Assistant Secretary of the British Treasury, as financial adviser to the Qajar government. Armitage-Smith appointed William McLintock to investigate Persian complaints against the government. Naturally the company considered that McLintock, being British, would uphold the company's claims that it had treated the Persians fairly in all circumstances.

McLintock's report was so honest that its details were suppressed by the British government. The details leaked out, however. McClintock's report verified all the objections listed above. Then he disclosed other ways that Anglo-Persian was cutting into rightful royalties due to Persia. One of these was a sinking fund, which the company had established to pay off its future bonded debts as they came due. All monies put into this fund, McLintock found, were immediately deducted as current expenses and charged against the royalties due to the Persian government.

The report put Anglo-Persian in such a bad light that it agreed to drop its claims against the Persian government for the war damage to the pipeline. It also agreed to pay $5 million to satisfy Persian claims and to pay royalties on each ton of oil instead of upon net profits. The company refused to guarantee to pay royalties on time, but did agree to pay interest when they were late.

These concessions were acceptable to the Persians, but they balked when the company flatly refused to consider giving up royalties on the company's operations outside Persia itself. The Persian government then angrily rejected the entire agreement.

That is where the matter stood when Reza Shah became dictator of the country. A few months after he came to power, Armitage-Smith was dismissed as financial adviser to the government, and Millspaugh, an American, was hired. Millspaugh succeeded in doubling the amount of royalties the government received, but was unable to change the inequities that caused these royalties to be far lower than they should have been.

Reza Shah and his prime minister, Qavam as-Saltane, now tried to involve the United States in the confused political maneuvering by offering Standard Oil Company of New Jersey concessions in northern Persia. These concessions would be in the provinces excluded from the D'Arcy concession held by Anglo-Persian.

The British objected strongly to giving the American company oil rights in the five northern provinces. Sir John Cadman, a director of Anglo-Persian, was sent to the United States to negotiate with Standard Oil. Cadman suggested that instead of fighting, which would cost both companies considerable money, they work together in developing the northern provinces. Standard Oil agreed to a 50-50 split with Anglo-Persian, but the Persians refused to extend any agreements with the hated British company.

Standard Oil then withdrew and Qavam began negotiations with the American-owned Sinclair Oil Company. The Sinclair deal featured a very fair royalty arrangement and called for the company to float a $10 million loan for the Persian government in the United States.

The deal collapsed. It was first delayed when a Persian mob killed an American vice-consul in Tehran when he tried to photograph a religious ceremony. It collapsed after Harry F. Sinclair became involved in the Teapot Dome scandal when it was revealed that his company had bribed the United States Secretary of the Interior to obtain leases in a naval oil reserve in Wyoming.

Reza Shah tried to reinterest Standard Oil in developing the five northern provinces, but failed because the Russians objected to an American company drilling so close to their southern border. They did not object during the negotiations with Sinclair because Sinclair had previously assisted Russia in a sale of badly needed petroleum products.

Anglo-Persian ignored Reza Shah's demands for increased royalties and the Persian dictator could do little about it. However, resentment about the company's attitude spilled over into other Middle East countries. They viewed the events in Persia as a forerunner of what could happen to them.

It is doubtful that Anglo-Persian's actions were illegal under the D'Arcy agreement. But this concession was drawn up by skillful lawyers who took advantage of the Persian government's lack of knowledge of the ways of oil-company negotiations. Thus, although the company's actions were not illegal according to the agreement, they were certainly morally wrong and showed complete insensitivity to Persian rights. The company followed a definite policy of finding every possible loophole to reduce Persia's share in oil revenues.

The actions of Anglo-Persian in the Middle East were certainly better than those of Rockefeller's Standard Oil Company during the founding years of the first great oil trust. It must be remembered that then, as now, big international concerns operated under the rule that their primary duty was to make the most profits for their shareholders.

In the light of history, Anglo-Persian's operations were on a par or may have been a little better than their competitor's. However, the company must take the blame for creating resentments that directly led to the present situation in the Middle East as it relates to oil.

The Persians, or Iranians, as they are now known, have little reason to sympathize with the Western world's agonizing energy problems. It they had been treated more fairly in the beginning, the entire Middle East oil situation might well be different today.

6

Reza Shah Fights Back

Reza Shah Pahlavi, king of kings of Persia, could scarcely read and write, but he was a leader of tremendous ability in spite of all that. He ruled his country with an iron hand, forcing it to shake off the shackles of the past and meet the challenge of the twentieth century. At the same time he developed pride in his people and a growing nationalism that included violent hatred of exploitation of the country by the British oil concern.

Although balked in his drive to get more of the oil profits for his country, Reza Shah never stopped trying. He noted that Indians imported to work in the refinery at Abadan were paid more than Persians for similar work. He resented it when the company curtailed production to raise world oil prices in the 1929 Depression. He received evidence that company officials were interfering in local politics in southern Persia. He also complained that the company was doing nothing to train Persians for higher positions.

His first new attack on the company was an income-tax law passed in 1930. The company refused to pay. It claimed that the royalty agreement was in lieu of taxes. There was no way Reza

Shah could collect, short of defeating Great Britain in a war. He nursed his rage and sought other means to break the company's stranglehold on Persian oil.

The Shah's next tactic was to demand a complete revision of the D'Arcy concession. The demand included 25 percent ownership of any company doing business in Persia, an equal share in any subsidiary anywhere in the world, reduction of the concession area under the D'Arcy agreement, reduction of the sixty-year concession to twenty years, and a forty-cents-a-ton royalty on all oil taken from Persian fields.

Opposition in Persia was now so great that the oil company became alarmed. The board of directors feared that uprisings might interfere with production. Again Sir John Cadman was sent to Tehran to talk with the Shah.

The company was also uneasy at rumors that the famous C. S. Gulbenkian was urging the Shah to cancel the D'Arcy concession and give it to a consortium of American and French oil companies. Gulbenkian was an Armenian who played a leading role in petroleum negotiations. He never owned a company, but handled deals for a percentage of stock or a fee. This was always 5 percent, and he became known as Mr. Five Percent. When Gulbenkian got into the act it was time for the opposition to worry.

The quarrel between the Shah and the oil company finally exploded into action after Anglo-Persian cut the royalty payment in 1931 to a third of what had been paid in 1930. The company insisted that this was necessary because the world Depression had lowered profits. The Shah noted that Persia's share of the oil revenue was only $1.5 million, whereas the British government collected $5 million just in taxes alone. In addition, the British government received profits on the 51 percent of Anglo-Persian stock it held. Persia got no taxes at all on the oil taken out of the country. It tried to collect them, but Anglo-Persian refused to pay.

The national assembly voted to refuse the royalty cut, but the assemblymen were only echoing the Shah, who made all major political decisions. In his next move Reza Shah ordered the controlled Persian press to print demands that Persia be given 40 percent of the oil produced.

Sir John Cadman indicated that Anglo-Persian might consider increasing Persia's royalty, but his offer was rejected as insufficient by Reza Shah. The Persian dictator visited the oil fields and returned more determined than ever to take the most drastic action to control what he considered his country's rightful heritage.

The climax of these years of bitter complaints came in November 1932. The Persian Ministry of Finance, at Reza Shah's orders, delivered a coldly formal note to the company's representatives. It set forth Persia's complaints against the Anglo-Persian Oil Company. It said that the original D'Arcy concession had been granted by a government (Qajar) no longer in existence and that Persia did not consider itself bound by it.

"The company has seen fit to reduce Persia's share of petroleum profits while increasing its own share," the note said. "Therefore, His Imperial Highness's government has no alternative but to annul the concession awarded in 1901." The note added that the Persian government was ready to grant the company a new concession on terms providing more justice to the Persian people.

The contents of the note were made public. There was jubilation in Tehran, and mobs thronged the streets shouting anti-British slogans. The note also caused an international sensation. Here was a political ant challenging a political elephant! In India, Mahatma Gandhi, who was also challenging the British government, watched to see what the outcome would be. There was also special interest in Mexico, where the government was similarly dissatisfied with arrangements they had made with American oil companies operating within their country.

The company naturally rejected the Persian note. Its position was that the D'Arcy agreement was a firm contract. It could not be amended or canceled until it ran its course—in 1961. The Majlis immediately passed a resolution reaffirming the note. The British government then stepped into the argument. The ambassador to Persia handed the Persian foreign office a note, which said in strong language that His Majesty's government would take all necessary steps to protect its interests in Persia.

British warships were immediately dispatched to the Persian Gulf to demonstrate the willingness of the government to support this declaration with force.

Reza Shah stuck to his position. His determination was supported by urgent and secret communications from his ambassador in London. These reports said that anti-government forces in the House of Commons were disturbed by the threat of war. The Conservatives favored taking over the Persian fields by force, but the majority of the Labor Party rejected such a move.

The Conservatives were in power, but their majority was narrow. The government decided that it could not risk a military invasion in the face of both Labor Party and world disapproval. There was no denying that the royalty arrangements and some of Anglo-Persian's money juggling were unfair to Persia. Instead, the British government notified Reza Shah that the British claim would be referred to the International Court of Justice in The Hague, the Netherlands, if the Shah did not withdraw his announcement of contract cancellation.

The British position was that, fair or not, the concession was a binding legal contract. Both government and company officials, believing in the sanctity of business contracts, felt that the world court would support them.

Reza Shah's position was that the oil dispute was a purely internal Persian matter and that the International Court of Justice did not have jurisdiction over disputes between a nation and companies working within its borders. He accused the British government of threatening Persia. He claimed this was a threat to world peace that he intended to refer to the League of Nations.

The British government moved first. It abandoned the idea of asking the International Court of Justice to hear the case. Instead, a complaint was lodged with the League, claiming that the annulment of the D'Arcy concession posed a distinct threat to British subjects working in Persia and thus threatened world peace.

Reza Shah was outraged at the way the British politicians had stolen his own plan and complained to the League first. His ambassador to The Hague then laid Persia's complaint before the world body.

The League, like the United Nations that succeeded it, was reluctant to take any kind of direct action. It asked the British and Persians to take no step that might endanger peace in the Middle East. Then after some delay the League appointed a commission, headed by Dr. Edward Beneš of Czechoslovakia, to look into the complaints.

The Persian delegation detailed their complaints against Anglo-Persian. The company's delegation vigorously defended its actions. It strongly pointed out that there would be no Persian oil except for the enormous risks and expenditure of capital poured into the project.

After each side had made its presentation and each had had the opportunity to rebut, Beneš adjourned the session to consider the testimony. At the same time he pressured both sides to settle the trouble between themselves.

The company sent Sir John Cadman back to Tehran in March 1933. This time Reza Shah swept aside all underlings. Sir John had to negotiate directly with the iron-hard Shah. Together they worked out a deal that both the company and the Shah ratified on April 29. The agreement was reported to the League, and the British complaint and the Persian cross-complaint were removed from the League's agenda.

The settlement called for a new concession that was to run for sixty years, ending in 1993. The government agreed that the concession could not be annulled or canceled during this period. In turn, the company fixed royalty payments at four shillings per ton of petroleum sold. The Shah was pleased with the raise in royalty, but the consensus in the oil industry was that Cadman had outfoxed the Persian ruler, because the royalty was to be paid on

Reza Shah, father of the present Shah of Iran, was the first Middle East ruler to force an international oil company to increase oil royalties.

Photograph of an oil painting in Gulistan Palace, Tehran, Iran

petroleum "sold." Thus, the company could avoid payments on oil pumped and stored, either in Persia or in any of the companies' refineries or tank farms around the world.

The Shah agreed to keep company operations free of taxes in consideration of an annual payment in addition to royalties on oil sold. Anglo-Persian agreed that this payment would never be less than £300,000 ($1.5 million) for a pound sterling was then worth about five American dollars. In all, the oil industry considered that Reza Shah had done better than any of them expected him to do, for they had not expected the British government to give in even this much.

The entire agreement ran through twenty-seven articles. There were numerous minor concessions that made Reza Shah feel that he had won a considerable victory. One of these, reflecting his concern at the low number of Persians in managerial positions, pledged the company to spend $50,000 a year to train Persians in England. Also, when the concession ended, all company property in Persia would revert to the government.

The agreement (the entire text is included as an appendix in Norman Kemp's *Abadan*) does not appear to be as grossly unfair as Reza Shah's son, the present Shah of Iran, would later claim. Nor does it appear to be as great a victory for the company as others have claimed.

Regardless of who got the best of whom in the deal, this concession agreement was earthshaking in regard to world petroleum production, and was another one of the direct causes of the present Middle East squeeze on world oil prices.

Something that no one seems to have noticed at the time would have a far-reaching effect very shortly. This point, which would hurt the oil industry so much, was contained in the preamble to the concession agreement. The preamble read:

> For the purpose of establishing a new Concession to replace that which was granted in 1901 to William Knox D'Arcy, the present Concession is granted by the Persian

Government and accepted by the Anglo-Persian Oil Company Limited. This Concession shall regulate in the future the relations between the two parties mentioned above.

The important phrase in this statement is "new Concession." From the beginning of the quarrel, Anglo-Persian and its supporting British government claimed that Persia had no legal right to cancel the old concession. The concession was a contract and not subject to cancellation as long as its terms were fulfilled.

In referring the matter to the International Court of Justice at The Hague, the company had followed the usual procedure when a contract is violated. It went to court.

Contracts are not as sacred as some people seem to believe. They can be overturned by courts. This may happen when the contract was obtained by fraud or when one side fails to live up to the agreement. A contract can be overturned when it is an "odious burden"—that is, the terms are so unfair as to create an unnecessary hardship on one of the parties.

If the matter had gone to the world court, the terms of the contract might have been revised under the "odious burden" doctrine. In any case, the alterations to the concession (contract) or its cancellation would have been due to court action.

Instead, the matter never was settled by the court. The Persian government unilaterally, or by its own actions, canceled the concession. In agreeing to a *new* concession rather than insisting just on a revision of the *old* concession, the Anglo-Persian Oil Company in effect acknowledged the right of Persia to cancel the previous agreement. Then, when the League of Nations accepted the settlement and removed the complaint from its agenda, it in effect put the stamp of approval on the concept that a nation did have the right to cancel unilaterally a concession given to a foreign company.

This was carefully noted by a number of small countries whose natural resources were being exploited by big nations. In the past they could do nothing about changing unfair concessions. Gunboat diplomacy of the nineteenth century would have involved them in instant war. Now the conscience of the world was moving away from imperialism. People were still sick of the bloody carnage of World War I, as both the French and British government found when they tried to back Middle East demands with army bayonets.

One of the first countries to follow Reza Shah's lead in demanding more control of its natural resources was Mexico. The Mexican government had been having trouble with international oil barons for some time. Huge petroleum deposits were discovered in Mexico in 1910. The president, General Porfirio Díaz, favored British oil companies. This brought about friction between British and American oil companies, and according to Anton Mohr, to a revolution that overthrew President Díaz in 1911, after thirty-five years as Mexico's dictator. Mohr claimed, "Indeed, the revolution was actually financed and promoted by American oil companies."

Mohr, writing in 1926, goes on to claim that the "revolutionary intrigues" of 1913 were due to British oil interests counterattacking. Americans complained that General Huerta, the new dictator, was in British pay. Unfortunately for those engaged in these oil politics, General Carranza, who overthrew Huerta, hated both sides. He forced through strong antiforeign laws that greatly hampered the oil companies. It was not surprising that as soon as World War I ended, three revolutionary groups, two financed by foreign money, attacked Carranza. (Pancho Villa, the third revolutionary, financed his battles by banditry.)

Succeeding presidents—all put in office by revolution—wanted more oil money for Mexico's treasury and, in some notable cases, their own pockets. The restrictive laws remained, although the more important ones were ignored by the oil companies. Finally, after seeing Reza Shah defy Great Britain in the Middle East, the Mexican government nationalized the Mexican oil industry, completely taking over ownership of the companies. The companies had been about 70 percent American-owned and 30 percent British-owned.

Reza Shah had not gone as far as nationalization, but he now considered it. He had grown increasingly oppressive as he aged, and had developed into a tyrant feared by everyone, friend and foe. He stayed in power only because of his control of the army, and he needed more and more money to keep his supporters in line. In an unindustrialized country like Persia, oil was his only means of raising additional funds.

After the Mexican nationalization, Reza Shah talked with his various advisers about the possibility of Persia's running its own oil industry. He discovered that there were not enough trained Persian oil engineers to operate the refinery if he threw the British out. The old British policy of barring natives from higher positions had proved to be good insurance. Reza Shah abandoned the idea of nationalization, but took some actions which indicated that the idea still remained as a future possibility.

For one thing, Reza Shah hated Britain because of the oil disputes and the occupation of Persia in World War I. He was equally distrustful of Russia, which was supporting Communist groups opposed to him. So he turned to Germany for technical assistance. All evidence indicates that the Shah was suspicious of Hitler's drive, but he needed help. German technicians were invited to come to Iran. At their peak there were four thousand of them in the country. The British viewed them with great suspicion.

The German technicians did not include petroleum engineers, but the Shah, considering future nationalization of the petroleum industry, hoped that he could depend upon Germany to furnish engineers to run the refineries if the British were forced out.

In the meantime, as a symbol of the new nation he was forging, Reza Shah renamed his country Iran, an ancient name going back to the days of Cyrus the Great, founder of the great Persian empire. He also demanded that the Anglo-Persian Oil Company change its name to the Anglo-Iranian Oil Company, to conform to the new national policy. The company obliged and became AIOC in the shorthand language of the oil industry.

Up to this point, Iranian oil dominated Middle East production, with Iraq second in importance. Then in the late 1930's a new factor entered into the churning politico-petroleum world. This was the discovery of oil in the Arabian Peninsula. These new fields eventually proved to be the world's largest petroleum reserves, ensuring that the Middle East would someday replace the United States as the major oil supplier of the world.

7
Arab Oil

In their rush to exploit oil in Iran (Persia) and Iraq, petroleum geologists had not overlooked possibilities in the vast deserts of the Arabian Peninsula. However, the region was split into many small sheikhdoms, each of which was constantly at war with the others. The Ottoman Empire controlled only a section of the west coast, including the religious centers at Mecca and Medina, and a shorter section of the east coast. The central and southern areas belonged to the belligerent sheikhs. Under these unsettled conditions, no oil company wanted to risk its capital trying to drill for oil.

The ancient history of the Arabian Peninsula is still pretty much a mystery. Arabian historians reckoned their history from the time of Muhammad, their great religious leader, and only in recent years have archaeologists dug in the peninsula. Their spades have turned up the ruins of great cities, proving that the early Arabs had a high degree of culture.

After Muhammad died in A.D. 632, his followers began a series of wars that spread their religion and enlarged their empire until it reached from India's Indus Valley across the Middle East into

Egypt and North Africa. The religion proved more durable than the empire. Later Arab conquests included Spain and half of France, but the empire had all disappeared by the eighteenth century, and the Arab peninsula was divided into independent sheikhdoms.

Then about 1703, a religious leader named Muhammad ibn Abd al-Wahab was born at Najd near Riyadh, the present capital of Saudi Arabia. This new Muhammad was extremely puritanical. He believed in strict interpretation of the Koran, the Islamic Bible. In time his followers became known as Wahabis, after their leader.

Like all religious reformers, Muhammad ibn Abd al-Wahab encountered strong opposition. He was fortunate to be supported by the Sheikh of ad-Diriyah (a town near Riyadh). The sheikh was a member of the Sa'ud family, who had ruled ad-Diriyah as long as men could remember. The Sa'uds protected al-Wahab and fought his battles. After he died they carried on his reform movement, building a gradually enlarging state as they defeated neighboring tribes.

By 1811, the Sa'uds were strong enough to make raids into Syria and to plan attacks on Baghdad. At this point the Ottoman Empire, which controlled the Middle East, sent an expedition against them. The Sa'uds were crushed and their leader taken to Constantinople for beheading. They continued strong in central Arabia, however, until 1871, when they were weakened by civil war between two brothers. This permitted the Turks to capture the east coastal area down as far as Qatar.

The Sa'ud family was in a bad way when its greatest member was born in 1880. They had to move constantly. When the boy—who became known as Ibn Saud—was eleven, the family fled to the edge of the Rub al-Khali (the "Empty Quarter"), the desolate desert in southern Arabia. Later they had to flee to Qatar and then to the island of Bahrain before locating in Kuwait near the head of the Persian Gulf. Here the Kuwait ruler, Mubarak al-Sabah, took a liking to the boy. Ibn Saud fought in Mubarak's battles, gaining considerable renown as a leader. In 1902, Ibn Saud was twenty-two, Mubarak supplied the young man with arms, and Ibn Saud set out to reconquer Riyadh, from which his family had been expelled years before.

For several months the tiny force maneuvered on the fringes of the desolate Rub al-Khali desert. Both Mubarak and Ibn Saud's father sent word for the young man to return to Kuwait. He continued to find excuses not to do so. Finally, in January 1902, Ibn Saud made his move. He left twenty of his sixty men with his camels a day and a half's march from Riyadh. The rest went on foot to an oasis outside the city. Here he left the remainder of his forces, except for six men who entered the city with him during the late hours of January 15.

After a careful reconnaissance, Ibn Saud sent one of the men back to bring the thirty-three men left at the oasis. When they arrived after midnight, Ibn Saud led them in an attack on the governor's house. They captured the governor's wife and sister, but the governor himself was sleeping in the city fortress and the fortress gates were locked for the night. It would have taken a far greater number of men to storm the gates successfully.

From servants captured in the governor's house, Ibn Saud learned that the fortress gate would be unlocked at dawn and the governor would return to his house. The young leader decided to wait for dawn and try to capture the governor.

The forty men spent the rest of the night in the house and said their morning prayer before taking their predetermined stations. Four men were placed on the roof to cover the area between the house and the nearby fort with their guns. The rest remained hidden behind the garden wall.

When the governor came out, accompanied by ten men, Ibn Saud rushed at him. The governor tried to cut his attacker down with a sword. Ibn Saud shot and wounded the governor, who tried to run back inside the fort. The defenders were unable to close the gate in time, and the forty attacking Arabs got inside. In a brief battle they defeated the eighty defenders, and killed the governor.

After taking possession of the city, Ibn Saud brought his father, Abd ar-Rahman, from Kuwait to govern while he set out to bring the rest of the Najd, or central plateau of Arabia, back under the family control. Ibn Saud's adversary, Ibn Rashid, got Turkish help and attacked Ibn Saud with artillery in 1904. Victory in this battle brought Ibn Saud great prestige.

Ibn Saud's domain originally was restricted to the central part of the Arabian peninsula. As the years passed, he extended his sheikhdom to the east coast and then in 1921 finally defeated the House of Rashid, which still controlled the northern section of the peninsula. This brought Sultan (as he was then called) Ibn Saud's domain to the border of Kuwait. Sir Percy Cox (who had been political officer in South Iran during Reynolds' time) acted as mediator to draw a borderline between Kuwait and Ibn Saud's domain.

At first Ibn Saud objected to a definite line between the two countries, because nomads, following the seasonal water and grass, passed back and forth through the area and he did not want a political barrier to stop their movement. Cox got around this objection by persuading the two rulers to set up a neutral zone between the two countries. This neutral zone is still in existence today. It required special treatment when oil was discovered there some years later.

In 1925 Ibn Saud captured Hijaz province on the west coast of the Arabian Peninsula, the region that includes the Muslim holy cities of Mecca and Medina. This completed Ibn Saud's conquests and he named his united kingdom Saudi Arabia.

Petroleum geologists had long believed that rich oil beds could be found in the Arabian Peninsula. Some adventurous souls had penetrated into the desolate area looking for favorable geological formations, but financiers were reluctant to invest money in Arabian explorations because of the political instability there. The end of the Arabian wars in 1925 caused them to take a renewed interest in the region.

The real pioneer of Arabian oil exploration was a New Zealander, Major Frank Holmes, an agent for Eastern and General Syndicate. This company had no intention of drilling for oil itself. Its purpose was to obtain drilling concessions and then to sell them to oil companies for a profit.

Holmes went first to Kuwait. Sandwiched in between Iraq and Saudi Arabia, this small principality seemed a good bet because of its nearness to the rich Iran fields. The sheikh, Hamad al-Khalifa, was suspicious of Holmes' interest in his land. He pleaded lack of knowledge of business and said he would consult Sir Percy Cox.

Holmes, impatient at the delay, went to Saudi Arabia. Ibn Saud listened but did not commit himself. However, he did let Holmes cross the peninsula on a personal exploration trip to seek possible mineral sites. When Holmes returned from the arduous trip, Ibn Saud gave him a concession to drill in a 30,000-square-mile area on the east coast of the peninsula.

As soon as this concession was signed, Holmes went to Bahrain, an island principality off the Arabian coast above Qatar Peninsula, and signed a drilling concession there in 1925. Holmes had done his part, but the directors of Eastern and General Syndicate back in London were unable to do theirs. No oil company was interested in buying the Arabian concession.

That was because their geologists had rendered unfavorable reports. Iranian oil had been found in porous limestone formations laid down in the Tertiary geological period, which extended from 60 million to 12 million years ago. Erosion, geological upheavals, and disturbances had exposed rock of this era in the Arabian concession area. Any oil present in these rocks would have evaporated or drained away. Since there was at that time no indication that oil might be found in deeper deposits of rock, the oil companies declined to buy the Arabian concession.

Eastern and General Syndicate then allowed the Arabian concession to expire. However, they did interest Gulf Oil Company, an American firm, in the Bahrain concession.

Gulf geologists returned favorable reports on oil prospects on Bahrain Island, but the concession sale hit a snag because Gulf was a partner in the Turkish Petroleum Company (later the Iraq Petroleum Company). The partners had made an agreement that none of them would seek oil concessions within a certain area of the Middle East without the consent of the other partners. This was called the "Red Line Agreement" because the area included had been circled by a red pencil line on a map. Bahrain was within the Red Line area.

The other partners refused permission for Gulf to proceed alone, declining to join in a partnership in an area where they lacked faith in petroleum prospects. Gulf then offered to transfer its agreement with Eastern and General Syndicate to the Standard Oil Company of California (Socal, in the shorthand language of the oil industry). Socal agreed and Gulf's option was transferred in December 1928. Socal then discovered that the British government, trying to forestall American penetration into an area it considered its own, had already taken steps to block competition.

Bahrain, consisting of one large and several small islands, lies about twenty miles off the al-Hassa province coast of Arabia. It was then extremely poor, with most of the population centered in the capital city of Al Manamah. The land was too desolate for much agriculture, and the chief industry, as it had been since Roman days, was pearling. Rich beds of pearl oysters were to be found in its surrounding waters.

The British, long years before, had established a protectorate over Bahrain. This was part of the British political policy of establishing its influence in the Persian Gulf region to protect the British trade routes through the Suez Canal to India. Under the terms of the treaty between the Sheikh of Bahrain and Great Britain, mineral concessions could be ceded only to a British company.

However, British financiers had shown no interest in opening oil drilling in Bahrain. The sheikh, desperate for money but afraid to disturb his relations with Great Britain, told Socal representatives he would honor the concession transfer from Eastern and General Syndicate if the American company could overcome British objections. This was easily done. Socal formed a subsidiary company in Canada, naming it the Bahrain Petroleum Company, Ltd.

The British government looked upon this arrangement for what it really was—an attempt to get around the government policy. A considerable amount of political pressure was brought to bear, but the government was not in a strong position because none of the British companies wanted to drill on Bahrain. Finally, in August 1930, permission was given for Bahrain to approve the transfer of the Eastern and General concession to the Socal-owned Bahrain Petroleum Company.

In the meantime, the Shell Oil Company had acquired a concession to drill in Asir province on the west coast of Arabia below Mecca. Drilling began in 1927, but was abandoned in 1929 after a series of dry holes. The failure of Shell—a predominantly British-owned company—helped convince the government that Arabia was not a likely source of future oil. One of the legends of the oil industry is that the government then agreed to permit Socal's Bahrain Petroleum Company (Bapco) to drill solely because the British politicians were sure that no oil would be found.

Drilling rigs, pipe, boilers, and crews had to be brought in. Land had to be prepared, water provided, and quarters erected for the men. It was October 1931, before drilling began on Bahrain. Bapco geologists had mapped an anticline and the first well was located directly in the center of the domed underground area.

The drilling was watched closely by all oil companies. Previous oil deposits had been found in the Middle East only in Tertiary rock deposits. Tertiary rock was exposed in the Arabian Peninsula, and any oil they might have held had long since disappeared. The Number 1 Bahrain well would be probing older rock deposits, reaching back to the Mesozoic period—the time of the dinosaurs.

The drilling progressed slowly, reaching 1,000 and then 1,500 feet in depth without any sign of commercial oil deposits. Then in May 1932, after almost eight months of drilling, the bit broke through into an oil pool at a depth of almost 2,500 feet—slightly short of a half mile straight down.

The oil pool proved to be contained in a porous limestone bed laid down in the Cretaceous age—the last period of the Mesozoic era, when dinosaurs became extinct and the first mammals appeared. Geologists give its approximate time as 120 million years ago.

Additional drilling during the next two years verified the extent of the oil field and revived other oil companies' interest in the Arabian Peninsula. Bapco found itself with oil, but no market. Neither Bapco nor its parent, Standard Oil of California, had an eastern marketing organization. So Standard made an agreement to form a joint company with the Texas Company (later renamed Texaco), which did have marketing facilities. This company was called Caltex (California-Texas Company). As part of the deal, Texaco became a 50-50 partner with Standard Oil of California in Bapco.

Industry interest now turned to Saudi Arabia. Canny Ibn Saud was no longer in a hurry to let oil companies come into his land. Reza Shah's troubles with the British in Iran made the Saudi Arabian ruler wary of the political implications of allowing foreign companies to operate in his country.

However, Ibn Saud had been closely watching operations in Bahrain. Spies kept him informed of American actions there. He was favorably impressed by the way the Americans kept out of local politics, in contrast to British actions in Iran and Iraq.

In 1931, an American engineer, Karl Twitchell, made an east-west survey for Ibn Saud to check on mineral possibilities. His report, like the one Holmes made earlier, was favorable, but nothing came of it until after the Bahrain strike. Ibn Saud, badly in need of money, listened when Standard Oil of California sent Lloyd Hamilton to see him in 1933. They worked out a concession agreement, but the deal was jolted when President Franklin D. Roosevelt closed United States banks and prohibited the export of gold as an emergency measure to combat the deep American Depression. The agreement with Ibn Saud called for payment in gold.

The concession now had to be renegotiated. It was finally

signed on May 29, 1933. Standard Oil of California formed a new company, California Arabian Standard Oil Company (Casoc), to do the exploration work. The company's name was changed to the present Arabian American Oil Company in 1944 and has been called Aramco ever since.

There was no need for new oil fields when Casoc petroleum geologists set out to find oil prospects in Saudi Arabia. The entire world was locked in a great Depression. Oil was selling for ten cents a barrel in the United States, and there was a plentiful supply.

However, the old companies were looking ahead to the end of the Depression. They also were concerned about the depletion of American oil. An enormous amount of United States oil had been supplied to Europe during World War I, and oil economists could already foresee a serious drop in United States production in the coming years. So even though oil was not needed in 1933, it certainly would be needed in the future. American companies wanted to ensure that they had a share of Middle East oil so they could stay in business if and when American production was insufficient to supply United States and their other customers' demands.

The men Casoc sent out to find oil in Saudi Arabia in 1933 were true pioneers—hardy souls who rode camels into desolate waste-lands, climbed rocky precipices on foot, and sweated and went thirsty in furnace-like heat.

Petroleum geology had come a long way since D'Arcy's men picked drilling sites on the basis of oil seepages. Seventy-four years had passed since Colonel Drake drilled the first oil well. Geologists now understood a lot about the underground conditions where oil was likely to be found. However, they still relied to a large extent upon surface indications. What they could see of an area usually gave some idea of what was underneath. Unfortunately, a large section of Arabia was covered with sand that masked these telltale indications.

Casoc had 320,000 square miles in which to find oil—and only eight men in the initial crew to find it. They were aided in 1934 when the company sent them an airplane. This lightened some of the drudgery of covering ground on foot and by camel. Now the geologists could inspect hundreds of square miles a day. Then, back at the base camp, they could pour over aerial photographs and verify their sightings.

Of course, promising sites still had to be inspected on the ground. In some hard-surface areas and salt beds they could bounce along in a car with a steaming radiator, but many places were impossible to get to except by camel caravan.

The first promising site was mapped at Dammam, a small fishing town on the east coast about twenty-five miles or so across the water from Bahrain Island. Although the site was near the proven Bahrain field, Casoc engineers and geologists did not rely on that. There was a dome formation at Dammam, which experience had shown was highly favorable to the collection of an oil pool.

Any oil man will tell you that geology can only tell you where oil may be found, for oil, like gold, is "where you find it." The only way to tell positively if oil is present is to drill a hole and find out. In this case, the Dammam dome was similar in geological structure to the Bahrain dome, which was producing oil. The company considered that good enough evidence of oil to risk drilling.

The decision was made in September 1934. The time necessary to bring in drilling equipment and build a camp for the drilling crew took the rest of the year and the next spring. The well was finally "spudded in"—drilling begun—in April 1935.

There were some long faces at Casoc headquarters when the results came. The faces got longer when wells Number 2, 3, 4, 5, and 6 also failed to produce oil in commercial quantities. They did produce some oil and a little gas, but nothing that warranted opening up a field.

Disappointed, but still undaunted, Casoc tried again. Well Number 7 was drilled to 2,500 feet, the depth of the Bahrain field, without success. The geologists advised going deeper. The dome formation in which they were drilling should have produced oil. By February 1938, they were down to 4,000 feet. Again the decision was made to go deeper.

By this time the oil industry had long since switched from the pounding of cable-tool drilling to the faster rotary drill. In this method a drill stem was rotated by a flat rotary wheel while water and mud were constantly pumped into and out of the hole to remove the chewed-up rock.

Drilling on Number 7 Well continued into March 1938, when the drill bit finally broke through the covering shale into a limestone dome laid down in the Jurassic period, about 35 million years older than the Cretaceous limestone in which the Bahrain oil was found. The well came in a roaring gusher. Later drillings set the limits of the pool at about four miles in length and at least three miles in width.

Jubilant oil prospectors, now assured that there was oil in Arabia, redoubled their efforts to find other fields. The problem was the miles on miles of sand that covered the telltale surface formations that geologists sought. One promising area, the Rub al-Khali ("The Empty Quarter"), is a desert of sand that stretches for several hundred miles across the southern part of the Arabian Peninsula. It is the largest stretch of pure sand to be found anywhere in the world. The wind piles up dunes that tower as high as 500 to 800 feet in the air—truly mountains of sand.

The Rub al-Khali is trackless. The constantly shifting sand obliterates all landmarks, so that wanderers crossing this vast dry sea must, like sailors, guide themselves by compass, sun, and stars. Forbidding, waterless, and often swept by devastating sandstorms as it was, still oil prospectors rode their camels into the Rub al-Khali in their never-ending search for oil.

Originally they used magnetometers, which measure the earth's magnetic field variations, and the gravity meter, which measures variations in gravity. From these measured differences, petroleum geologists could get a rough idea of underlying rock strata, but

A wildcat well put down by Aramco drillers in the 1940's in the Empty Quarter of the Arabian Peninsula, where sand dunes rise as high as five hundred feet.

Courtesy of Aramco

such methods, though helpful, could not provide all the information the oil prospectors needed.

In 1937 the geologists began using the newly developed seismographic system. In this system, holes are dug to predetermined depths and high-explosive charges are set off in the holes. The shock waves from the explosions jar down through the underlying rock and then bounce back to where seismographs (instruments designed to measure the shock waves of earthquakes) record them. The waves reflect in different strength from different rock strata, enabling geologists to obtain a more precise idea of the hidden layers of rock.

Helpful as it was, the use of the seismograph by these "doodlebug" crews, as they were called, could tell the geologists only that subsurface structures were *favorable* for finding oil. Only drilling would tell for sure. So new "wildcat" wells were drilled in unproven territory to test the geologists' educated guesses.

As exploration continued in Saudi Arabia, sheikhs controlling the numerous small sheikhdoms along the lower eastern and southern coasts naturally became interested in oil as a means of raising their own low standard of living. Oil explorers from various companies were invited into Qatar, Oman, Yemen, Abu Dhabi, and Dufar.

Opening the discovery well on the Dammam dome in Saudi Arabia did not immediately bring great wealth either to Ibn Saud's government or to Casoc. It takes a long time and a tremendous investment of money to develop an oil field. Years ago someone said that the people who develop oil fields are neither businessmen nor geologists. They are gamblers who prefer to shun the easy way to lose money like shooting craps or playing stud poker. They like to do it the hard way by hunting little pools of oil securely hidden inside a great big world.

Those who complain about oil prices and oil company profits have no idea of the enormous capital and time that goes into developing a field. In the case of the Dammam dome field in Arabia, seven years elapsed from the time Casoc signed the concession agreement with Ibn Saud until the first oil was pumped into a tanker for export.

During these years, Casoc paid out millions in expenses with absolutely nothing coming in from Saudi Arabian fields. Payments began with a $100,000 payment to Ibn Saud to sign the agreement. To this were added other agreements such as road-building and other aids to the government, which raised the initial payments to well over a million dollars. Next came the costs of exploration, including salaries, equipment, facilities, and often bribes to local tribesmen to prevent trouble. There was less of the latter in Saudi Arabia under Ibn Saud's tight reign than in Persia when Reynolds was having so much trouble.

The cost of making an exploratory drilling rose from about $20,000 a well to as much as $1 million, depending upon the depth and the trouble encountered. In the case of the famous Rig 20 fire in Persia, when a new well exploded in flames as it came in, a famous oil fire fighter had to be flown in from the United States to fight the fire for weeks before it was brought under control. Direct and indirect costs, including loss of oil, probably ran $2 million or more as a result of that fire alone. Today, sinking a well offshore in seawater may cost as much as $20 million.

In the Dammam dome field in 1937, the discovery of oil was only the beginning of even greater expenses for the company. New wells were sunk to define the extent of the pool. A refinery had to be built at Ras Tanura on the coast above the field. Loading docks had to be constructed to ship the oil. The wildcatters' camp at Dammam exploded in size to accommodate the influx of workers and grew into the present Arabian city of Dhahran. New highways had to be constructed to handle the trucks servicing the field. Pipelines had to be laid, storage tank farms erected, and garages built to service the equipment. Also adding to the enormous cost of developing the initial Arabian Peninsula field was the fact that drilling equipment and supplies had to be shipped eleven thousand miles from the United States.

The first oil from the Arabian fields began flowing into a tanker at Ras Tanura on May 1, 1939. King Ibn Saud came from his capital, Riyadh, for the historic occasion. The king was accompanied by a retinue of two thousand including seventeen members of the royal family and four hundred soldier bodyguards. Ibn Saud toured Dhahran and the oil fields. He then went to the port of Ras Tanura, where he opened the valve to permit the first oil to flow into the tanker. The refinery at Ras Tanura was not yet in operation. The tanker was loaded with crude oil, which is oil as it comes from the ground, except in this case some of the poisonous sulfur dioxide had been removed at Dhahran.

In time the Saudi Arabian and other peninsular fields would prove to contain the world's largest reserves of petroleum.

Casoc teamed up with the Texas Company, as it had in Bahrain, in order to get access to a worldwide marketing system. Then, in 1944, the company changed its name from California Arabian Standard Oil Company (Casoc) to the Arabian American Oil Company, and immediately became known as Aramco. In the beginning Aramco was owned by Standard Oil of California and the Texas Company (Texaco), but later some other major United States oil companies would share in the new company.

Tragic political events were piling up to an explosion the year Ibn Saud opened the valves to start Arabian oil flowing to market. Within four months World War II began, with its tremendous effect upon the oil industries and oil politics in the Middle East. The events that followed proved once again that Middle East oil and international politics were inseparable.

8

Invasions of Iraq and Iran

After World War I, Lord Curzon, the British statesman, had said that "the Allies floated to victory on a wave of oil." Oil was even more important in World War II with its greater use of air power and mechanized weapons. When the war began in September 1939, the Middle East, with its tremendous oil reserves, became a region of primary concern to both sides. Neither Great Britain nor Germany had any oil fields within their own territory, so it was inevitable that Great Britain would take whatever means it considered necessary to preserve its stake in Middle East oil. At the same time Germany would be looking for an opportunity to seize the oil lands that Adolf Hitler's war machine needed so badly.

After the fall of France, the German occupation of Greece, and the entrance of Italy into the war on Germany's side, Hitler could spare troops from the other fronts, and was in a good position to attack the Middle East oil fields.

When the expected attack came it was not made by Germany directly, but through Iraqi rebel generals. At the end of World War I Iraq had become a British mandate under the League of

Nations. Amir Faisal became king in 1921 with the British still in control. Continued Iraqi opposition to British mandate control caused Faisal to ask in 1922 that the mandate be changed to a treaty between Iraq and Britain. Finally in 1927 Great Britain agreed to recognize Iraq as an independent country, but won the right to keep air bases and troops there to protect British oil interests. The mandate, however, was not officially terminated until 1932.

When World War II began in September 1939, the Iraqi government was headed by General Nuri es-Said, who favored the British, but he was deposed in March 1940, by a military leader, Rashid Ali al-Gailani, an anti-British nationalist.

The new government was admittedly pro-German, which posed a definite threat to British oil interests in the rich Kirkuk fields. British troops were immediately moved to Basra, the Iraqi port where Kirkuk oil was loaded on tankers.

Al-Gailani ordered the Iraqi army to drive the invaders out. When the fighting started on May 2, 1941, other Iraqi army units took possession of all British oil fields and refineries. Foreign-born employees were put under house arrest, and in some isolated areas a few were put in concentration camps. The oil wells were not sabotaged, but company stores and offices were looted. The Iraq Petroleum Company had shown more consideration for the local people than Anglo-Iranian had done. As a result, there was little bitterness toward the British and the interned British employees were not harshly treated in most cases.

The fighting ended late in May. Al-Gailani fled to Germany and Nuri es-Said, the pro-British former premier, was reinstated as head of the government.

Al-Gailani had expected direct German aid against the British, and German spies actively encouraged him in this belief. Otherwise, he would not have dared pit his small army against strong British troops. There was considerable mystery as to why German aid was not forthcoming, but the mystery ended on June 22, 1941, about three weeks after the Iraqis surrendered. Hitler had never intended to furnish direct military aid to Iraq. He was planning a surprise attack on Russia. A major German objective was to capture the Russian oil fields at Baku near the Caspian Sea, and Hitler had wanted the Iraq revolt to tie up British troops on a new front.

The German attack on Russia created a tense political situation in the Middle East. Winston Churchill, the British prime minister, warned Russia's premier, Joseph Stalin, that German troops in occupied Greece could easily push through neutral Turkey and move into northern Iran (Persia), where they could then attack southern Russia. Stalin, worried about the German drive on Moscow, was not immediately concerned with Churchill's worries.

Churchill persisted, arguing that Russia and Great Britain should jointly invade Iran. This would protect Russia's ''back door'' and British oil production in the south. Stalin still held back, reluctant to pull needed troops from the German front.

Reza Shah, the Iranian ruler, had for some years encouraged German technicians to come to Iran. He needed technicians to help with his broad policy of modernizing his country, and Germans were among the best. Also, he feared Russia and Britain. Both had invaded Persia during World War I, and he felt that good relations with Germany, a rising power, would act as a dampener on future Russian and British aggression. There were 650 German technicians in Iran. (British intelligence reports blew up the number to over 4,000.)

When Germany invaded Russia in June 1941, Reza Shah repeated his declaration of neutrality. Stalin then sent a diplomatic note warning Iran to order German nationals out of the country. Reza Shah agreed, but stalled on carrying out the order. Other Russian and British protests followed. The Shah continued to stall. This encouraged the British belief that the Iranian ruler was pro-German and that the German technicians were a ''fifth column'' awaiting an opportunity to sabotage the oil fields and refineries. Also, those British connected with the oil industry still smarted over the way Reza Shah had highhandedly—in their opinion—upset the D'Arcy agreement and forced through royalty changes in 1933. They argued that such a man could not be trusted.

Immediately following the German attack on Russia in 1941, President Franklin D. Roosevelt announced that the United States would extend lend-lease aid to Russia in its battle against the invading Nazis. The big problem was how to get these munitions and supplies to Russia. The direct sea route to Murmansk was extremely hazardous because of German submarines. The safest way was to ship the matériel into the Persian Gulf. Then it could be carried over the Iranian railroad to the Caspian Sea, where Russian freighters could pick it up.

The need to use Iran as a supply transfer point decided Stalin to go along with Churchill's demands for a joint invasion of Iran. It is possible that permission to transship supplies through Iran could have been obtained from Reza Shah without an invasion. His son, Mohammed Reza, the present Shah of Iran, insists that this is so. However, it must be remembered that at the time Germany was definitely ahead in the war. France had fallen. The British had suffered a great defeat at Dunkirk, and the Russians were falling back under the first shock of the invasion of their territory. It is unlikely that Reza Shah would have risked Hitler's displeasure by siding with the Allies at that particular time.

In any event, Russia and Great Britain did not bother to suggest negotiations. On the morning of August 24, 1941, they jointly invaded Iran. The Russians moved in from the north. The British attacked from the Persian Gulf, striking hard to secure the oil fields and the Abadan refinery. The fighting was stiff for five days, ending with an armistice on August 29, 1941.

Under the terms of the cease-fire, Russia occupied the five northern provinces. Great Britain occupied the area surrounding the oil fields. Tehran, the capital, was not occupied, and the occupying powers pledged themselves not to interfere in Iran's internal affairs—a pledge that neither Russia nor Great Britain kept.

Reza Shah was told that he was no longer acceptable as ruler of the country. He agreed, because after being an absolute monarch he was totally unable to accept orders from the Russian and British generals heading the occupation troops.

The Shah left the meeting with the occupying generals and went to see his twenty-one-year-old son. "I am abdicating in your favor," he said. He showed no anger or bitterness at his enemies. "All I can tell you by way of advice is to put your trust in Allah and never be afraid."

The old man then went into exile on a British ship that took him eventually to South Africa, where he died in 1944.

Ever since he was six years old, Mohammed Reza, the new Shah, had been trained by his father to be the next king of ancient Persia. Now that the time had come, he found himself more a prisoner than a monarch. Two thirds of his country was controlled by foreign troops. Local political groups were fighting for power. The Tudeh party—a Communist-supported organization—was building a power base from which to depose the Shah when the war ended. The country's economy was a shambles and inflation was causing wide distress.

It is a tribute to the new Shah's extraordinary ability that he was able to keep his throne under these conditions. In time he was able to raise Iran from a defeated, impoverished land to its present position as the dominant country in the Middle East.

Oil and skillful politics provided the key to the Shah's successes, but a long, bitter fight lay ahead before he could secure control of his country's oil. He rightly considered oil to be the only means he had to revive Iranian world importance.

The war drastically increased the use of oil, but caused a huge drop in Iranian production. In 1938, the year before World War II began, Iranian wells had produced 10 million tons of petroleum. This dropped to 6.6 million tons in 1941, reducing Iranian royalties at a time when war inflation was booming other prices. The reason for the drop was that Iranian oil was marketed primarily in

The Shah and his first wife, Queen Fawzia, visit the radio station serving American soldiers assigned to the Persian Gulf Command in 1944.

Courtesy National Archives

Europe. Sales stopped as Hitler's forces overran country after country.

The depression in Iranian oil sales changed abruptly for the better when Japan attacked Pearl Harbor in December 1941, and the United States entered the war, for it cut off access to Far East oil. Then when Great Britain and the United States invaded North Africa in 1942 the demand for Iranian oil shot up even higher.

The refinery at Abadan was enlarged. Aviation gasoline became a priority product to support the invasion of Sicily in 1942 and the invasion of Italy that followed. Anglo-Iranian oil production jumped from its low of 6.6 million tons a year to 13.25 tons in 1944, when the second front opened in Europe with the cross-

Channel invasion of France. Production reached 17 million tons in 1945 before dropping off again when the war ended.

Anglo-Iranian had enjoyed a monopoly in Persian oil since the D'Arcy agreement was signed in 1901. Now both Russia and the United States tried to move in. In 1944 representatives of the American Sinclair Oil Company and Socony-Vacuum Company came to Iran. Separately they sought drilling concessions in the northern provinces not controlled by Anglo-Iranian.

This alarmed the British, who put political pressure on the Iranian prime minister to reject the American advances. The Shah also objected to leasing the area until after the war, when Iran would be in a better bargaining position. The American offer was turned down.

As soon as the Americans failed, a Russian delegation arrived in Tehran with a demand for rights to explore the five northern provinces excluded from the D'Arcy concession. The Shah rejected the Russian demands, although he knew it would cause diplomatic trouble with the Soviet Union. He was depending upon the British to restrain a Russian attack.

However, Russian troops began maneuvering in the affected areas. Then Communist mobs, led by members of the Tudeh party, stormed through the streets of Tehran. At the same time there was a wave of angry nationalism throughout the entire country against giving away any more of Iran's oil. An old-time politician, Mohammed Mossadeq, seized the opportunity to make himself the champion of those opposing foreign ownership of Iran's oil resources. Mossadeq cleverly exploited antiforeignism to the point where, for a time, he was the most powerful man in Iran and nearly dethroned the Shah.

At this point, however, the Shah was in hearty agreement with Mossadeq and those who opposed foreign ownership of Iranian oil concessions. The Shah had built grandiose dreams about restoring Iran's greatness. He needed tremendous sums of money to do this, and his only means of getting it was through the oil Anglo-Persian was taking out of Iran.

During these war years the Shah's hands were tied, but he was already looking forward to the period ahead when Russian and British troops would have left Iran. At the moment he felt safe in defying Russia because he did not feel that British and American interests would let Stalin take the northern Iranian provinces by force. He was proved right when both the British and the American ambassadors to Moscow presented their countries' objections to the Russians.

Russia then eased its pressure on Iran to grant it the northern oil concessions, but gave definite indications that it would consider it an unfriendly act if Iran tried to give these concessions to any other country. Meanwhile, Russia kept building up the Tudeh party's power in these provinces.

Russia, Great Britain, and the United States had all pledged to honor Iran's independence at the end of the war. Troops were to be completely withdrawn within six months under an agreement signed by Russia, Britain, and Iran. Now the Shah nervously eyed the growing strength of the Russian-backed Tudeh party. The party had grown so strong that it seemed possible that Russia could honor its agreement to withdraw occupation troops from northern Iran and still hold the northern provinces through revolutionary action by the Tudeh.

As the Shah struggled with his petroleum-inspired political problems in Iran, the situation was entirely different in Saudi Arabia. Leaders of the American oil industry realized that the United States would eventually lose its dominant position in world petroleum production. This country had supplied 6 billion barrels of oil to the Allied World War II war machine. New discoveries of American oil were not keeping pace with the amount of petroleum being consumed. It was clear at this point that within a few years the Middle East would replace the United States as the major supplier of world oil.

The United States had a foothold in Middle East oil through the American-owned Aramco with its concessions in Saudi Arabia. However, the American government had never made any effort to develop ties with Ibn Saud. Saudi Arabia had received some American lend-lease aid, but it had been extended through the British.

In 1944 President Franklin D. Roosevelt, spurred by warnings from American oil interests, realized that this action was cementing Saudi and British relations, and could result in British oil companies' obtaining concessions in Saudi Arabia also, thus competing with Casoc's plans for future expansion. As a result, Cordell Hull, the American Secretary of State, arranged that future aid to Saudi Arabia would go directly to Ibn Saud through the United States Department of State. This would ensure that the Arabian ruler knew who his benefactor was.

It was later denied that petroleum considerations were behind this move. However, Cordell Hull verified it in his *Memoirs*, published in 1948. At the same time, Harold Ickes, Secretary of the Interior, drew up a plan whereby the United States government would buy controlling stock in Casoc in the same way that Winston Churchill had arranged for the British government to buy control of Anglo-Persian in 1914.

Anglo-Persian had welcomed the British government as a partner because the directors realized the advantage this gave the company in international politics related to the oil business. The American oil industry—rugged individualists all in those days— wanted no United States government interference in their business operations, although they welcomed the government's "good offices" when political pressure was needed. In a word, they wanted the government as a supporter, but not as a partner. Influential oil figures discussed the matter with President Roosevelt, and the Ickes plan died a quiet death. Shortly after this, in January 1944, Casoc's name was changed to Aramco.

Later, when Aramco's big expansion required enormous amounts of new capital and marketing facilities, the original owners, Standard Oil Company of California and Texaco, Inc., took in Standard Oil of New Jersey and Mobil Oil Corporation, as additional partners.

Exploration had continued in Saudi Arabia during the early war years. The country had the oil, but not the means of exploiting the markets. A plan was developed whereby the United States government would finance a pipeline to the Mediterranean coast at a cost of up to $160 million. Aramco would be given twenty-five years to repay the cost and would guarantee to maintain sufficient oil reserves for United States supplemental military requirements.

The plan fell through because both the oil industry and the United States Senate disliked the idea of the government's getting into private business. Instead it was decided that the small refinery Casoc had built at Ras Tanura would be enlarged by Aramco, using its own funds. However, the American government would provide top priorities so that steel could be bought for the project. The government also aided the company by providing military transport for some equipment and personnel.

This intense activity in the Saudi Arabian fields helped develop them faster than would have been possible without government support. At the beginning of the war in 1939 there was just the start of an oil industry in Saudi Arabia. When the war ended there were four major fields and a greatly expanded refining capability so that Saudi Arabia could immediately expand into a major producing country. Ibn Saud could thank politics—and British-American jealousy—for turning his country, practically overnight as oil industries are developed, into a major petroleum producer.

In 1944, when the big push began, Saudi Arabia was producing less than 20,000 barrels of oil a day. By 1947 this had increased twelvefold, and then jumped to half a million barrels a day by 1949. It was only the beginning of what later became a fabulous production.

Other future oil-producing sheikhdoms of the Arabian Peninsula were not so fortunate. Oil had been discovered in Qatar in 1939, but the wells were capped for the duration of the war. Similarly, plans to prospect for oil in Oman, Abu Dhabi, Yemen, and what became the United Arab Emirates were postponed. Development in Kuwait also was slowed by the war. These

countries had to sit back and jealously watch Ibn Saud grow rich
while they still maintained their bare subsistence living.

The ending of the war caused a sudden drop in the oil production that had fed the war machine, but it was considered only a
temporary recession. The gigantic task of rebuilding a devastated
Europe would require a lot of oil. Thus, Saudi Arabia and Bahrain
were in an excellent position to cash in on the new oil market
while the emirates were struggling to build an oil industry.

Saudi Arabia's position was improved by the construction of
Tapline, a pipeline to carry Saudi Arabian oil to the Mediterranean coast. Since the majority of Arabian oil was marketed in
Europe, the pipeline could move the petroleum much cheaper than
it could be freighted by tankers. Tapline began in July 1945, when
Aramco formed the Trans-Arabian Pipe Line Company, whose
initials became the shorthand name for the company. It took two
years to work out transit permissions and royalties with Syria,
Jordan, and Lebanon, through whose territories the pipeline
would extend. Actual construction did not begin until 1947.

Thirty-seven years had passed since determined men from
D'Arcy's company built the first pipeline for Middle East oil in
Iran. Now huge ditching and earth-moving equipment were available to ease the labor. The pipe joints were welded together
instead of being joined by steel collars that had to be twisted on by
human labor. But despite the advances the years had brought,
building Tapline was in its own way as heroic an achievement as
the work of the pioneers.

The actual construction work was done for Aramco by Bechtel
Corporation of San Francisco, California. It began with a monu-
mental job of assembling American skilled welders, pipe fitters,
machinists, and engineers who were willing to fight sand and
120-degree heat to push Tapline across three nations and 1,068
miles from the Persian Gulf to the Mediterranean Sea.

The first base camp was set up in July 1947 at Ras al Mishaab.
The rest of the year was spent assembling equipment (numbering
over seven thousand different items), training Arab laborers, and

building a rock jetty where pipe could be unloaded. The pipe was 30 and 31 inches in diameter. The different sizes were chosen so that the smaller pipe could be placed inside the larger and thus take up only half as much space aboardship. This cut transportation costs because the ships charged by the amount of space taken up on board rather than by weight.

The joints of pipe were 31 feet long. Welders in the base camp joined three of these together to make lengths of 93 feet. This work went on while engineers fought their way across the desert in trucks and cars as they surveyed the area the pipeline was to pass through, or its right-of-way.

The actual Tapline construction began at Qaisumah, east of Ras al Mishaab. Gathering lines were constructed to bring oil from the various fields to this central point, where Tapline would take it on across Saudi Arabia, Jordan, and Syria to Sidon in Lebanon. Tapline proper consisted of 753.5 miles of pipe with 314.7 miles of gathering line from the fields to Qaisumah.

Pipe began to move from Ras al Mishaab in February 1948. The 93-foot lengths were hauled to the right-of-way on huge truck-trailer units. One load of pipe weighed as much as forty tons. Tractor lifters unloaded the pipe and strung it along the right-of-way. These sections were then joined by the use of automatic welding equipment—the first time automatic welders were used for field operations. Since the joints had already been welded into 93-foot lengths in the base camp, the number of welds that had to be made along the desert right-of-way was reduced, greatly speeding up operations.

As the huge pipe trucks dumped their loads and the welding crews melted the joints together, leveling and ditching crews sweated and toiled ahead of them to prepare the way. Where possible the pipe was buried in ditches but, in some areas that could not be done. Then they followed the lead of the first Iranian pipeliners who had just laid the pipe across the top of the ground. However, they improved on Ritchie's ground lines for D'Arcy by suspending the line above the ground, using rings set in concrete.

The pipe used for this purpose had to be especially designed to allow for the stresses set up by the contraction and expansion that occurred between the very low night temperatures and the 120-degree heat of the day.

The route of Tapline, beginning at Qaisumah, south of the Iraq-Kuwait neutral zone, followed roughly the border between Iraq and Saudi Arabia to the Jordan border. It then crossed northern Jordan, proceeded along the Syrian border, and crossed Lebanon to the ancient Phoenician port of Sidon on the Mediterranean coast. The final weld was made on September 25, 1950, and the first tanker was loaded at Sidon with Tapline oil on December 2.

In the course of construction, prime pump centers were set at distances of about 175 miles along the pipeline, with automatic pump stations in between. Portable industrial X-ray equipment was dragged along the construction route to check each weld to make sure it was sound to prevent future breakdowns.

While postwar oil operations in the Arabian Peninsula ran smoothly from the beginning, the situation in Iran was different. As soon as the war was over in Europe in May 1945, the Shah sent diplomatic messages to Russia and Britain reminding them that under the Tripartite Treaty they had signed with Iran they were bound to remove their occupying troops from his country within six months after the war ended.

Neither Britain nor Russia showed much inclination to end the Iranian occupation. When the Shah became more insistent, Great Britain honored its commitment under the Tripartite Treaty and British troops were removed from Iran. Russia still ignored the Shah's demands. It became apparent that Russia was determined to hold on to the northern Iranian provinces, either through continued occupation or by setting up a puppet republic of these five provinces under Tudeh rule.

Then the Tudeh announced that it was forming a National Assembly, which would rule the region as an autonomous—self-governing—part of Iran. They stopped short of declaring total independence for the northern provinces. The Shah immediately order the Iranian army to put down the revolt. The Russian army of occupation stopped the Iranians at the border of the occupied territory.

The Shah turned to the newly formed United Nations. On January 19, 1946, he filed the first complaint received by the world body, and the UN Security Council received it with considerable reluctance. Russia was infuriated at being the subject of the first complaint, and the other big powers were unhappy at being forced into a showdown with Stalin at this critical time when the United Nations was just beginning.

The small nations of the world watched the proceedings with tremendous interest, for here was a small nation asking for justice in a dispute with a world power. The action of the United Nations in this complaint would show whether or not the world body would protect the rights of small nations. If it did not do so, the United Nations would be wrecked. On the other hand, if it did take the Iranian complaint, Russia might withdraw from the United Nations, which would also weaken the organization.

The Security Council sought some kind of compromise instead of standing up to its obligations. On the suggestion of Jan Christian Smuts of South Africa, the Security Council skirted the issue by suggesting that Russia and Iran hold further discussions.

The discussions were carried on by Iran's pro-Russian premier, Ahmed Qavam. The Shah was unhappy with Qavam but was not strong enough to control him. However, when Qavam gave Russia an oil concession in the provinces, agreed to autonomous government in those areas, and allowed Russia to leave troops there, the Shah exploded. He instructed his ambassador in the United States to demand that the Security Council consider his complaint.

At this point President Harry S. Truman became concerned about the future effect this precedent might have on the United Nations. He sent what has been described as practically an ul-

timatum to Stalin to honor his previous agreement to withdraw
Russian troops from Iran. Shortly afterward Andrei Gromyko, the
Russian delegate to the United Nations, announced that Russia
would withdraw from Iran.

Iran's troubles with Russia and the Tudeh did not end there, but
since this book is the story of oil in the Middle East, the international politics and troubles of the countries have no place here
except as they bear on petroleum production. Unfortunately, very
serious trouble related to the oil industry was just ahead for the
Shah and his country. For a while it threatened to destroy the
Iranian oil industry and the Anglo-Iranian Oil company as the
Shah, Anglo-Iranian, the British government, and a strange, often
weeping fanatic named Mossadeq fought a savage three-way
battle for the future of Iranian petroleum.

9
Mossadeq and Abadan

Mohammed Mossadeq was a Persian aristocrat. He was born in 1881 into a family with large feudal land holdings in Iran. As was fashionable for sons of wealthy families, Mossadeq was sent to school in Europe. He then went into government service and rose rapidly. He was elected to the Majlis—the Iranian assembly—in 1915, and was reelected each year for the remainder of the Qajar rule. He was extremely ambitious and often became so emotionally involved in his oratory that he fainted.

In the revolution of 1921 he supported Sayyid Zia al-Din, but secretly hoped to replace him. When Reza Shah made himself dictator and then Shah, the fanatic Mossadeq developed a vicious hatred for the new ruler. Finally, in 1938, Reza Shah grew tired of Mossadeq's opposition and had the aging politician jailed. Crown Prince Mohammed Reza—the present Shah of Iran—persuaded his father to release Mossadeq. Years later Mohammed Reza wryly admitted that this had been one of his most serious mistakes.

Although Mohammed Reza had succeeded in getting Mossadeq released from political prison, the older man felt no gratitude. He

hated the son as much as he had hated the father. After Reza Shah
was forced to abdicate by the British and Russians at the beginning of the occupation during World War II, Mossadeq began
building a political party aimed at overthrowing the monarchy.
This eventually became the Iranian National Front party.

After Mohammed Reza had the Crown estates—his personal
feudal lands—divided among the peasants in 1950, Mossadeq
mounted his most ambitious campaign to drive the young ruler
from the throne. In this effort Mossadeq had the full support of
both the Tudeh Communists and the wealthy landowners. It was a
strange combination of political bedfellows, but they opposed the
Shah more than they opposed each other. Both knew that any
cooperation between them was simply a matter of convenience. It
would end when they broke the Shah.

In the meantime, the Shah developed a seven-year plan for his
country, but he lacked money to put it into effect. As his father
had done in 1932, Mohammed Reza looked to Anglo-Iranian as
the source of the needed funds. He wanted more royalty money.

The company firmly rejected the suggestion to change the 1933
agreement. Company lawyers quoted the final paragraph of Arti-
cle 20 of this agreement: "This Concession shall not be annulled
by the Government and the terms therein contained shall not be
altered either by a general or special legislation in the future, or by
administrative measures or any acts whatever of the executive
authorities."

Powerful political forces immediately began hate campaigns
against the British company. They accused the company of keeping Iran in "beggary" while the company "wallowed in wealth"
stolen from the Iranians. The Tudeh party formed labor unions
and began crippling strikes in the oil fields and refineries. Mossadeq seized on the oil dispute to inflame the people against both
the company and the Shah, whom Mossadeq blamed for his
"inability to force the British to pay a fair price for Iranian oil."

The company made extravagant claims about the fairness of its
treatment of Iranian employees. However, except for those in

higher positions, the laboring class was poorly paid. They did not have company housing and were forced to live in hovels. Postwar inflation made things even worse.

Even the pro-company British journalist Norman Kemp, who wrote a blow-by-blow description of the British expulsion from Abadan, admitted that housing was badly substandard. He wrote, "In contrast, thousands of Iranian workers were crammed into inadequate dwellings. In spite of its efforts to stem the mounting flood of people without decent homes, the Company could not keep pace." He claimed it would take an entire generation for adequate housing to be erected for all the workers, and blamed the shortage on lack of trained carpenters and materials.

The first trouble broke out in the Agha Jari field, a location that had been abandoned in 1938 after several dry holes were drilled. Under the pressure of war needs, drilling had resumed and an important new oil pool had been discovered. Just a week after the war ended in Europe in May 1945, ten thousand workers— stimulated by Tudeh agitators—went on strike.

The strikers demanded increased ice and water rations and a special pay allowance for working in the blisteringly hot, remote site. At one point overzealous company officials cut off all water to the strikers, who had no other source except company wells. The officials intended to force them back to work, but their plan backfired for religious reasons. Muslims are required by their religious laws to cleanse themselves as part of the ritual of preparing for their five-times-daily prayers. Although they are allowed to cleanse themselves with sand if water is not available the strikers looked upon the company action as a direct attempt to hinder them in practicing their religion. Fortunately the company realized the explosive situation they were encouraging, and restored water to the strikers.

While the strike was in progress, the Iranian Majlis passed a comprehensive labor law to govern minimum wages, working hours, overtime, holidays, and settlement of disputes. The failure of Anglo-Iranian to observe the new law led to a strike of six thousand men at Abadan on July 2. The company agreed to all demands except the one that required full pay for Friday holidays. Friday is the Muslim sabbath.

The strike was called off on government orders, but the radical press, controlled by the Tudeh party, accused government officials of collaborating with the British to oppress the workers in the oil fields. Wilder accusations followed. The British were charged with inciting the local sheikhs to withdraw from Iran and form a British-protected state. Another charge was that Iranians were prevented from holding managerial jobs with the company and that Iran was being robbed by the foreigners.

When the company refused to give in on the Friday pay dispute, the Tudeh led a general strike beginning at 6 A.M. on July 14. Rioting broke out in the late afternoon, resulting in forty-seven deaths. British warships were immediately put on the alert. Two ships moved up the Persian Gulf toward Abadan, but before they arrived on July 17, the strike had ended. The Iranian minister of commerce, Mozaffar Furuz, personally came to Abadan. In no uncertain language he informed the Tudeh-led union that the strike was illegal under the new labor law. He also informed the company that they would also have to conform with the new law, which included the Friday holiday with pay.

The strike had ended, but it was only the beginning of a long series of strikes, disturbances, and attacks by the anti-British factions in the government and of violent attacks by the Tudeh. The Tudeh attacks came to an abrupt end in 1949 when an assassin, disguised as a news photographer, shot and wounded the young Shah. This brought an instant crackdown on the Communist party, although, actually, no one ever knew who was behind the would-be assassin, for he was shot to death by the Shah's bodyguard. Reports that the dead man had a Tudeh party card in his pocket do not seem to have been true. Also, a later attempt on the Shah's life turned out to have been inspired by a fanatical right-wing religious group.

In any event, the Tudeh got the blame for the attack on the Shah and were forced underground. Elimination of the Tudeh did not lessen the violent attacks on Anglo-Iranian, however. Mossadeq, using the growing oil crisis as his road to power, increased his furious denunciation of the government for its failure to bring the giant oil company to terms. His wild speeches whipped the nation into a frenzy of hatred against foreign oil interests. He preached that all Iran's money problems could be ended if the country controlled its own Allah-given petroleum.

Anglo-Iranian finally realized that the situation was getting beyond control. The fact could not be disguised that Iraq was getting three times the royalty per ton of oil that Iran was getting. Saudi Arabia was sharing a 50–50 arrangement with Aramco. Also, despite the company's claims, Iranians working for Anglo-Iranian were treated as second-class citizens.

L. P. Ewell-Sutton, who was in Iran during much of the disturbance, wrote later in his book *Persian Oil*:

> The general level of wages in the company was always very low. . . . even after the general raise forced upon it by the Persian government in 1946. . . . When responsible members of management spoke contemptuously of Persians and denied them admittance to their homes, members of management paid lip-service to Persian susceptibilities in their official capacity and openly flounted them as private individuals, it is hardly surprising that newly arrived members of the staff automatically followed suit.

The company, through its officials and the British ambassador in Tehran, threatened the Iranians with hints of British army intervention. The Shah was in a difficult position. Mossadeq was denouncing him as a tool of the British, but actually he had little power at this time. The Shah's actions in dividing the Crown estates among its tenants and his refusal to bow to the Russians had increased his personal popularity with the masses. However, he was intimidated by the tremendous support the people were giving Mossadeq.

The Shah knew that the aging fanatic wanted to create a republic and drive him from his throne, but he did not feel strong enough to oppose Mossadeq. In most cases, the Shah disagreed with Mossadeq, but when the British threatened to send troops into Iran to put down the strikes in the oil fields, he threw his full support behind the old man. He told the British ambassador that he would personally lead the troops into battle if the British invaded, and threatened to take the matter to the United Nations Security Council.

Company officials considered the situation. They did not feel that the British people would support an invasion of Iran just to protect the oil company's rights, even if the British government was willing. Instead, the company developed the Supplemental Agreement of 1949. This called for a 50 percent increase in royalties, stipulated that 20 percent of the money annually placed in the company's reserves should be paid immediately to Iran, and reduced the price of petroleum products sold in Iran. There were additional concessions that made the agreement a very fair one as far as financial arrangements went.

The agreement was signed by a government official and a representative of the company, but a Majlis revolt, led by Mossadeq, kept the agreement from being ratified. The historian Stephen Longrigg, author of *Oil in the Middle East,* the definitive history of the petroleum industry in that area, claimed Mossadeq's opposition was for "tactical political reasons almost completely unconnected with oil."

The Majlis' term expired and a new assembly and a new prime minister took office in February 1950. Ali Razmara, supported by the Shah and the Iranian army, was in favor of the Supplemental Agreement, but he could do nothing against the violent objections of Mossadeq and his National Front party. In fact, support for Mossadeq was so great that Razmara had to withdraw his objections to the delay in approving the Supplemental Agreement.

In the meantime, Hussein Maki, a member of the Majlis from the oil districts and a supporter of Mossadeq, went to Abadan, where he led opposition and agitation against British operation of the Iranian oil industry. Norman Kemp, who was there, described the situation: "Rabble-rousers organized demonstrations, with poets to recite heroic verse which were incentives to revolt, and launched slanging denunciations against the British. . . . Masses would sway ecstatically. They jumped, brandishing their arms frantically and vied to cheer the loudest."

The situation was getting completely out of hand again, both in the oil fields, where strikes were in progress, and in Tehran, where mobs were denouncing the British. The British ambassador, alarmed at the fury, sent a note to the Iranian government complaining that the people were not aware of all that Great Britain had done for them.

> I feel therefore, [the diplomatic note said] that the time has come when a strong line of conduct should be adopted by the Iranian government in order to explain what has already been done to help Iran, and to make sure that future discussions on the subject of oil should be conducted in a reasonable spirit and in full knowledge of the facts.
>
> I fear that the apparent willingness of Iranians to permit themselves uninstructed criticism of the Oil Company and of Great Britain is founded upon past prejudices which take no account of the immense service to mankind in general of the British people in recent times.
>
> It is, in my mind, most regrettable that public opinion in Iran should, as is apparently the case, cling to the out-of-date conception of England as a Power anxious to impose imperialism and colonialism wherever it can.

After receiving the British note, Prime Minister Razmara got up the courage to speak out against Mossadeq's demand for nationalization of the Anglo-Iranian assets in Iran. Razmara said that Iran did not have the ability to run the fields and refineries without foreign help. The Shah agreed with him. Mossadeq replied with a torrent of abuse upon both men.

A few days later, on March 7, 1951, Razmara paid for his

opposition to Mossadeq when he was killed by an assassin's bullet. There was a hasty conference in high Iranian military councils. The question of armed intervention and the jailing of Mossadeq and other National Front leaders was discussed. This was reluctantly rejected because Mossadeq had too much popular support, which included lower-ranking members of the armed forces. The generals did not believe they could win at this time in an armed showdown with the nationalist fanatic.

The Shah, who has been described as extremely indecisive during this critical period, agreed with the generals. In excusing his failure to confront Mossadeq, the Shah said later that he did not feel he could act until he was sure he had popular support. At that time he did not have it.

The political crisis continued. The revived Tudeh supported Mossadeq, and he was able to force an oil nationalization bill through the Majlis. The chaotic situation in Iran was closely watched by oil-producing nations throughout the world. Many were interested in bringing an end to foreign domination of their petroleum industries and were eager to see how Iran made out. It was true that in 1918 Russia had been the first country to nationalize its oil industry, but that situation had been different. It had been a Communist takeover and *all* industry had been nationalized. The Mexican oil expropriation in 1937 was also different. The nationalization in that instance was a takeover of foreign-owned assets only.

Under the circumstances there would be strong pressure for the British government to intervene with troops. In fact, immediately after the nationalization act was passed, British troops were moved into Cyprus, from where they could be quickly flown to Iran. An additional warship was also dispatched to the Persian Gulf.

The crisis caused the fall of Iranian Premier Hossein Ala's government, due to pressure from Mossadeq. By this time the vindicative old man had forced the resignation of five prime ministers and was accused of conniving in the murder of a sixth.

The Iranian constitution provides that the Shah shall designate the next prime minister, who must then form a cabinet that can survive a vote of confidence or no confidence by the Majlis. The Shah realized that no government could survive without Mossadeq's support. With considerable reluctance he asked the eighty-one-year-old politician to form a cabinet. Mossadeq accepted on April 29, 1951. The Majlis, dominated by Mossadeq's National Front party, ratified the appointment with a 99–3 vote of confidence.

The Anglo-Iranian directors realized that they could not compromise with stubborn old Mossadeq. Rioting and strikes increased at Abadan. The British cruiser *Mauritius* anchored off the island. The British government said the ship was there only to evacuate British personnel if the situation got worse. Mossadeq loudly denounced the ship's presence as "gunboat diplomacy."

Diplomatic observers expected another British invasion, but Iran was protected by British politics. A Labour government was in power in London. It had nationalized several major British industries, and was hardly in a position to object to another country's doing exactly the same thing with its own natural resources.

Also, there was the ticklish question of what action Russia would take. Russia still coveted the northern Iranian provinces and had reluctantly withdrawn under pressure from President Truman. If Britain invaded the southern oil fields, it was very likely that Russia would move into the north. There was also the question of possible United Nations action.

As rioting and other disorders mounted in the oil fields, the company asked that the disagreement be settled through arbitration by a third party, as provided for by the 1933 concession agreement. Mossadeq refused. He said he would settle for nothing less than total British withdrawal. He claimed that the entire oil revenue was Iran's and that Anglo-Iranian had no claim to any of it. He announced that Iran would run the company and that former customers of Anglo-Iranian's distribution companies could buy directly from the new National Iranian Oil Company (NIOC), which he had organized.

After discarding the idea of a military takeover of the oil fields, the British turned to the International Court of Justice at The Hague. Mossadeq refused to defend the British charge, claiming that the dispute was an internal one between the government and a private company. There was no international question at stake.

As the political maneuvering was going on, strikes and disorders continued in the oil fields. Mossadeq sent officials from his newly formed National Iranian Oil Company to take over direction of the company. Oil tankers putting into Abadan were refused cargo unless they signed receipts to the NIOC instead of Anglo-Iranian. The British employees refused to work under the direction of NIOC officials and production ceased.

The International Court, expecting to take about a year to decide the complaint, asked that Iran take no further action until a decision was reached. Mossadeq ignored the order and went on with his plans to take over British holdings.

The British government sent the Stokes Mission with a plan of compromise, whereby Iran would take over the production end of the industry and Anglo-Iranian would handle worldwide marketing of the oil produced. (There is more profit in marketing than in production, since the capital investment is less.) Mossadeq refused. He demanded all or nothing, and in the end he got all—and almost ended with nothing.

Refining operations were shut down in late July, and on October 4, 1951, the last Englishmen left the Iranian oil fields and refineries. It was the end of an era. Fifty years had passed since D'Arcy secured his first concession, and forty-three years had gone by since Reynolds' dogged stubbornness brought in the first important oil strike in the Middle East.

A week before the British staff left Abadan, the British government asked the United Nations Security Council to intervene. The United Nations, as always, hesitated to take a firm stand. It was glad to agree to an Iranian request to postpone action until October 15 so that Mossadeq could personally attend the session.

Mossadeq made a very effective speech to the Security Council, and it was a personal triumph for him when the United Nations decided to wait for the International Court's ruling before taking further action.

Mossadeq returned to a hero's welcome in Iran. In Britain the Labour Party lost the election, returning Winston Churchill to power. Political maneuvering continued. The long-awaited International Court decision ruled that it had no jurisdiction in the case.

This decision was hailed as a tremendous triumph for Mossadeq. He had resigned as prime minister five days before this (on July 17, 1952) in a political move against the Shah, and his followers had immediately begun such rioting that the frightened Shah was forced to ask Mossadeq to form a new government on July 21. Then the International Court decision, announced the next day, made Mossadeq the most popular man in Iran. Mossadeq made it clear that the days of the monarchy were numbered, and Anglo-Iranian had to face the fact that its holdings in Iran were gone.

The next phase of the fight centered on the company's efforts to get compensation for the loss. At the same time, the company informed the world that until such time as adequate compensation was paid, it considered all oil exported from Iran to be the property of the Anglo-Iranian Oil Company. Any purchaser of such oil would be sued by Anglo-Iranian for receiving stolen property.

It was the purpose of the company to deny Iran any petroleum markets outside its own national boundaries. Anglo-Iranian backed up this threat by slapping a lien on a small Italian tanker that had loaded a cargo of Iranian oil and then stopped at Aden, a British protectorate on the southwestern tip of the Arabian Peninsula. Then on January 20, 1953, another Italian tanker was accused of shipping "stolen" oil, and a lawsuit was filed in an Italian court. This was followed by a charge that the National Iranian Oil Company had sold several cargoes of oil to Japan at a 50 percent discount.

Despite these so-called "outlaw" sales, distribution of Iranian oil fell off badly. The British threat had effectively dampened any chance for Iran to build an international market of its own. At this point Anglo-Iranian officials believed that the international boycott would force Mossadeq to come to terms. The Iranian economy was in bad trouble because of the loss of oil revenues.

Mossadeq ordered drastic cuts in government spending. He vowed that Iran would "beggar" itself before giving in to British "slavery" again. He felt that the world needed Iranian oil and that this would put pressure on Great Britain to lift the oil embargo.

Things did not work out as Mossadeq had expected. The boycott of Iranian oil was a heaven-sent opportunity for producers in the Arabian Peninsula. They all stepped up production. Kuwait, Saudi Arabia, Qatar, Bahrain, and the Emirates all profited hugely as their sales increased to make up for the lack of competing Iranian oil.

The Shah made no criticism of the way his fellow Muslims took advantage of Iran's difficulties, but he was known to be bitter about it. Years later, when the Arab countries boycotted the United States and the Netherlands because of their aid to Israel, the Shah refused to take part in the boycott. He still remembered how the Arab countries had refused to support Iran in its time of need.

President Harry Truman considered Iran an important bulwark against the spread of communism in Asia. As a result, he extended considerable foreign aid to Mossadeq. This American aid helped the Iranian government withstand the crippling effects of the British oil boycott.

British oil interests viewed Truman's aid to Iran as an unfriendly act, but the British government made no comment. Quite possibly oil politics were at work here also. The British had cut American oil companies out of Iraq. American companies could not be considered heartbroken because the British Anglo-Iranian Oil Company was in trouble now. It was hinted that American oil companies expected, through the foreign-aid generosity of their Congress, to gain a hand in Iranian oil when production resumed.

The political situation changed when Dwight D. Eisenhower became President in January 1953. Mossadeq immediately warned the new President that there was danger of Iran's going Communist if it did not receive more American aid.

Eisenhower replied (according to a State Department bulletin, dated July 20, 1953):

> The Government of the United States is not in a position to extend more aid to Iran or to purchase Iranian oil. . . . I note the concern reflected in your letter at the dangerous situation in Iran and sincerely hope that before it is too late that the Government of Iran will take such steps as are in its power to prevent a further deterioration of the situation.

Eisenhower's reply meant that the United States had lost faith in Mossadeq as a bulwark against communism. It had decided to back the Shah instead. The Communist Tudeh party was rapidly becoming the most powerful force in Iran, but Mossadeq delayed any crackdown on the Communists because he needed their pressure against the Shah.

What happened next is still shrouded in secrecy. However, enough has been revealed to prove that the American Central Intelligence Agency (CIA) was active in Iran at this time. Quentin Roosevelt, grandson of President Theodore Roosevelt, has been named in an investigation by the United States Congress as having been involved in CIA activity in Iran at that time. The Shah of Iran's official biographer admitted that the CIA was involved, but insisted that the coup that overthrew Mossadeq was carried out primarily by Iranians.

On the basis of information available today, it appears that CIA agents disguised as embassy officials made contact with General Fazlollah Zahedi, a former member of the Majlis, and assured him of United States support. Zahedi contacted dissident officers in the Iranian army. The CIA then helped plan the attack and used American money to buy convenient mobs, who stormed the streets demonstrating against Mossadeq as a prelude to Zahedi's military action.

There was a leak somewhere. Mossadeq learned of the coming attack. He immediately suspended the supreme court and disbanded the senate, where he did not have a majority as he did in the lower house, the Majlis. His followers began deserting him as economic conditions approached chaos and it became clear that the Shah at long last was moving against his old enemy.

The Shah signed a decree removing Mossadeq from office and nominating General Zahedi as the new prime minister. An army colonel, backed by a platoon of soldiers, was sent on August 13, 1953, to deliver the decree to Mossadeq.

The old man had been warned by Tudeh spies, and he had his own soldiers waiting. They surrounded the colonel's platoon and lodged the leader with the decree he carried in jail. Mossadeq then went to the Majlis, where he delivered a stinging attack in which he denounced the Shah as a tool of the hated British. Shouting and weeping, he proclaimed Iran a republic, and ordered troops to arrest the Shah. The Shah fled to Baghdad and then to Rome.

The Tudeh Communists, who had temporarily supported Mossadeq, now turned on him in an attempt to take over the government. Mossadeq's police battled Communist mobs throughout the city. At this point General Zahedi and his supporting army officers, with CIA backing, moved in at the same time that pro-Shah mobs—paid to agitate by the CIA—demonstrated in the streets. Within hours the royalists held the city. *The Saturday Evening Post,* in its issue of November 6, 1954, called the coup "another CIA-influenced triumph."

The Shah returned to Tehran on August 22. Zahedi became prime minister. Some expected him to replace the Shah, as the Shah's father had taken over after a similar military coup. However, the Shah was able to ease Zahedi out after a time and started his own climb to his present position of power.

The Shah was eager to resume oil production, both to relieve Iran's economic squeeze and to provide funds for modernizing the

Propane, butane, ethane, pentane, and hexanes are made from natural-gas liquids in the NIOC refinery outside the Persian Gulf city of Bandar-e-Mah, Iran.

country. However, Mossadeq's propaganda had so inflamed the country against the British that it was totally impossible to return the nationalized industry to Anglo-Iranian—even if the Shah had wished to do so, which he did not.

The situation in the oil industry had shown that Iran could not effectively operate its petroleum resources without foreign aid. However, the Shah made it clear as negotiations resumed that the return of the British was unacceptable. The stalemate was finally resolved by formation of a consortium to operate the fields and market Iranian petroleum. Anglo-Iranian was represented in the consortium, which also included American, French, and Dutch companies.

In a complicated arrangement the consortium formed two companies to operate in Iran. One was devoted to exploring for new fields. The other was to operate producing fields. These companies could not sell oil. All oil produced had to be sold to the National Iranian Oil Company. NIOC would then resell the oil—either crude or refined—to the consortium's trading companies, who would market it at such prices as they could get. Profits of the consortium were to be split 50-50 with Iran.

In a separate agreement, Iran pledged itself to pay £ 25 million (about $70 million) for Anglo-Iranian property in Iran. In addition, the consortium paid an initial sum and a 10-cents-a-barrel royalty to Anglo-Iranian until the company had received $510 million. This was to compensate the British company for the loss of concession areas that the consortium took over.

Oil started to flow into tankers one month after the consortium agreement was signed on October 29, 1954. Six million tons of oil were produced in the first year of renewed operations. This increased to 35.8 million tons within three years. The Shah lavishly used the inflow of wealth to build himself the strongest army in the Middle East—an army that has made him virtually an absolute monarch.

10
The Gathering Storm

Practically overnight Iran became the most powerful nation in the Middle East. This was a tremendous comeback for the weak, poverty-ridden nation that had had to endure the humiliation of foreign occupation only thirty years before. The secret of Iran's climb to international importance was control of its petroleum resources—a control that it gained by hard, bitter fighting.

The lesson in Iran's struggles was not lost on the rest of the Middle East oil-producing nations. The Iranian victory was the end of an era for the petroleum industry and the beginning of a new one in which "colonial" oil exploitation would gradually disappear as producing country after producing country regained control of its natural resources.

While all the Middle East countries coveted the profits made by the various companies holding concessions in their oil lands, none immediately wanted to risk the sacrifices Iran had made. They did not want their fields closed for three years because they lacked the trained personnel and marketing channels to keep the companies going if the concessionaires were dispossessed. They preferred to work more slowly. The same thing was true in other oil-producing countries throughout the world—in South America, Indonesia, and North Africa.

The next big push for national control came in Iraq. Operations there had been serene from the end of World War II in 1945 until 1958, when a revolution overthrew the monarchy. In 1948, as a gesture of Arab resentment at the formation of the State of Israel, the Iraq Petroleum Company closed its pipeline to Israel. It has never been reopened.

The new government in Iraq was a military dictatorship. All the former government officials who had worked so long with the oil company were deposed and their places taken by men who not only knew little about the petroleum industry, but who also put political considerations ahead of the welfare of the company.

Like all military dictatorships, the new regime needed money to maintain the large army necessary to control the country, and it looked to the oil revenues to provide it. The first demand came in 1960 when the government placed an additional dollar-a-ton duty on all crude oil shipped from Basra through the Persian Gulf. Oil pumped over the pipeline to Lebanon was not immediately affected.

The company argued that this duty would place Iraqi oil at a price disadvantage in a market that then had a surplus. The government refused to listen. As a result, sales dropped and the company had to cut back on production, which meant that the government got less money than before it increased the Basra duty. The government blamed the company for the loss of revenue.

At this time Iraqi petroleum production was monopolized by three companies: Basra Petroleum Company, Iraq Petroleum Company, and Mosul Petroleum Company. From 1960 on the companies faced increasing government interference. The Iraqi dictator, General Abdul Karim Qasim, became increasingly erratic, hitting the companies with unreasonable demands that kept changing constantly. In December 1961, the first major blow fell when Qasim announced what became known as Law 80. This law took away from the companies all their concession lands except those where production was actually going on.

Qasim also demanded all exploration maps, seismographic records, and geological data for the surrendered area. He then formed the Iraq National Oil Company (INOC). It appeared from this that he was patterning his strategy after Iran with a view toward nationalizing the Iraqi oil industry.

Qasim was assassinated in 1963, and the situation improved a little. However, the new government did not restore the lost concession areas taken from the companies by Qasim. Abdul Salam Aref became president, but later led a military coup that set him up as dictator. Where he was killed in a helicopter crash, his brother, Adbul Rahman Aref, seized power. Aref was displaced by still another coup in 1968, which brought former premier Ahmad Hasan al-Bakr to power.

Conflict continued between Iraq Petroleum and the government during these turbulent years. The basic cause was Qasim's cancellation of Iraq Petroleum's rights to new concession areas in northern Iraq. This cancellation kept Iraq Petroleum from expanding. It limited the company's operations to old wells and such new wells as it could drill in the old production fields.

Iraq offered other companies the northern concessions, but the companies refused, for fear of future nationalization. They did not trust the Iraqi government. This prevented any expansion of oil production.

The government, refusing to acknowledge its mistakes, blamed Iraq Petroleum. During the first quarter of 1972, production dropped 17 percent from the previous year. Al-Bakr then shut down Iraq Petroleum and took over operations. Mosul Petroleum and Basra Petroleum, also foreign-owned, continued under their owners for a time and then they, too, were nationalized.

The Cinderella country among the oil-producing states of the Middle East, during the years immediately following World War II, was Kuwait. This small nation, no larger than the state of

Connecticut, was a desolate wilderness jammed in between Iraq and Saudi Arabia at the northern end of the Persian Gulf. There was an almost total absence of water, and wells dug through the sand only produced brackish water too salty for human consumption. As a result, the country's meager population centered around Kuwait City on the Gulf coast. The people lived by pearl diving, boatbuilding, and smuggling goods into the territories of their more prosperous neighbors.

In 1923 Frank Holmes, who made the first oil exploration of Saudi Arabia, tried to tie up leases in Kuwait, but the suspicious Sheikh Ahmad turned him down. Later Holmes did secure leases in the Kuwait Neutral Zone, the area along the border of Kuwait and Saudi Arabia. Ibn Saud of Saudi Arabia had objected to drawing a firm border between the two countries because the local nomads wandered back and forth through the area, and it would handicap their movements to find grass and water if firm national boundaries were drawn. Under the agreement drawn between the two rulers, any oil discovered in the Neutral Zone would be equally shared by the two countries.

The discovery of oil in Bahrain and Saudi Arabia changed the sheikh's mind. In 1934 he granted a concession to the Kuwait Oil Company, which was jointly owned by Anglo-Iranian of England and the Gulf Oil Company of the United States. Kuwait Oil was organized solely as a production company. Its oil—if any was found—would be sold to its parent companies, which would do the marketing.

Drilling began in May 1936, but an 8,000-foot hole failed to strike oil sand. Finally in April 1938, at a depth of 3,675 feet, drillers penetrated what became the famous Burgan field. It proved so rich that a geologist was quoted as saying that Kuwait was nothing but a pile of sand atop a sea of oil.

As in the other oil-producing states, the discovery of oil did not immediately make Kuwait rich. The outbreak of World War II caused such material shortages that the Kuwait Oil Company had to cease operations until the war ended.

The end of the war made it possible for Kuwait to share in the great boom that affected all the Arab oil-producing countries. Writing of the extraordinary developments immediately following the reopening of the Kuwait oil fields in 1946, Longrigg said,

> The record of the Kuwait Oil Company, which without waste or disorder brought oil production in six years from nothing to a rate exceeding three million tons a month, and equipped a remote waterless desert with all the necessities and many of the luxuries of a civilized modern life, is sufficiently remarkable to interest the most jaded mind.

Oil development in Kuwait was a model for the industry during these years. To their mutual benefit, the government of Kuwait and the companies operating there got along fine, which had not been the case in either Iran or Iraq. In turn, the Kuwait government used its oil wealth for the benefit of the people. This quickly produced in Kuwait one of the highest standards of living to be found anywhere in the world.

The small size of Kuwait and its enormous petroleum reserves naturally attracted the envy of other nations. They were kept at bay by the British, who had exercised a protectorate over Kuwait for the past century.

Great Britain withdrew in 1961, and immediately Iraq's dictator, Abdul Karim Qasim, claimed that Kuwait traditionally belonged to Iraq. British troops moved back into Kuwait. Qasim decided against trying to occupy his oil-rich neighbor. In more recent years the Shah of Iran has guaranteed Kuwait's independence.

Kuwait, with less than a million population, had a surplus of funds even after establishing a welfare state. This extra money, after 1972, was used to aid Arab nationalism. Gifts and loans were made through the Kuwait Fund for Arab Economic Development to Morocco, Bahrain, Jordan, and Sudan.

Despite having more oil revenue than the country needed, Kuwait followed the rest of the Middle East oil-producing coun-

A team of tractors skid a drilling rig in Kuwait to a new location about 1948.

tries in demanding control of its natural resources. In 1972 Kuwait joined other members of the Organization of Arab Petroleum Exporting Countries (Saudi Arabia, Dahrain, Qatar, and Abu Dhabi) in demanding and receiving an immediate 25 percent share in oil-company management. The agreement called for the Arab countries to get an increasing share of management until it reached 51 percent in 1983.

The oil companies did not object to this agreement. They thought it an alternative to outright nationalization. It did not disturb immediate arrangements or call for extra payments, but merely required a participation of the governments in the actual management of the oil companies. It would, it is true, give the Arab governments management control in 1983, but this would not affect their profits—or so they thought.

However, after the agreement was made between the

negotiators, the Kuwait assembly refused to sign it and insisted upon a 60 percent immediate takeover of the Kuwait Oil Company from British Petroleum (formerly Anglo-Iranian) and Gulf Oil. The two companies were paid $56 million for this 60 percent interest.

The trend since World War II had been for the national governments to demand greater and greater participation in the profits and finally the management of the companies operating within their borders. Through nationalization and participation agreements they had gone about as far as they could in this direction, for they now—for the most part—were in control of production within their territories.

However, oil production is not the most profitable part of the petroleum industry. Drilling, pumping, and refining are expensive processes that require large investments. On the other hand, selling and distributing the oil require less capital investment and produce greater profits.

Ever since Iran tried to get a share of Anglo-Iranian's worldwide marketing profits, the Arab countries had considered how they could increase profits still further by controlling the worldwide price of oil. Competition had prevented this, but competition—as John D. Rockefeller had shown in his original Standard Oil trust—could be broken by forming a giant cartel that could dictate prices to consumers. The Middle East oil-producing nations now began to consider how they could apply Rockefeller's monopoly techniques to their own situation.

11
OPEC

The price of foreign oil is based upon a "posted price." This posted price was originally set by the oil company offering the petroleum for sale and was based upon the quality of the oil, the market demand, and the expenses of production. The posted price was FOB (free on board)—that is, it was the price of oil delivered at the port of origin. The buyer then had to tack on the cost of transportation, further refining if necessary, taxes in his own country, his profit, and his distribution costs. All of these items added together, plus the retailers' profits and expenses, determined the pump price of gasoline and the price of other petroleum products to the consumer.

The posted price was extremely important to the oil-producing countries, for their royalties were based upon it. From 1950 to 1957 the posted price of Arabian oil, delivered at Ras Tanura, rose from $1.71 per barrel to $2.08 per barrel. Kuwait oil, which is of a lower quality than Arabian oil, sold for $1.65 a barrel in 1950 and rose to $1.85 by 1957. However, at the same time, United States Gulf oil, delivered in Houston, Texas, rose from $2.67 a barrel in 1950 to $3.07 in 1957. The oil, of course, was of a higher gravity, and more valuable, but that did not account for the great difference in price. American oil costs more to produce.

Nevertheless, the Middle East oil-producing countries were pleased by the gradual rise in their royalty incomes. This situation came to a shocking halt after 1957, when overproduction and a drop in world demand began to lower oil prices. Arabian oil dropped from a high of $2.08 in 1957 to $1.80 in 1962.

The price of refined products showed an even greater drop. Gasoline with a 90-octane rating sold at Persian Gulf ports for $4.49 per barrel in 1957. The price dropped to $4.20 in 1960 and was down to $3.78 by 1963. This meant that gasoline that was selling at the pump for about 37 cents average in the United States was costing 9 cents per gallon average at the Persian Gulf port.

The oil-producing countries were naturally alarmed. The prices of materials they were buying from Europe and America, as the basis of their modernization programs, were rising at an inflation rate of about 5 percent a year. Petroleum was not affected by the rising prices of other commodities. Instead it was dropping, cutting Middle East and North African oil-producing countries' revenue while their expenses kept rising.

In a four-year period they had seen their actual revenues drop about 20 percent on the average, while the cost of the materials they had to buy increased about the same amount. Thus they were faced with what amounted to an actual 40 percent drop in their buying power.

There seemed to be nothing they could do about it. In a free-enterprise market, prices are set by supply and demand. There was an oversupply and a decreased demand. Therefore, the price came down.

This simple economic explanation did not sit well with the oil-producing countries. The trouble—for the oil companies and the oil-consuming countries—was that too many Arabs, Iranians, and Libyans had attended capitalistic schools like the Harvard School of Business Administration. They understood the realities of the free-enterprise system just as well as those who tried to explain the revenue drop to them. In fact, as later events proved, they may have understood the realities of the business world better than their Western opponents.

What the oil-producing-country economists understood was that prices based upon supply and demand could be kept high if the seller could "corner the market." That is, if the oil producers could get together and create a monopoly, they could set their own prices.

In the past it had not been possible for the various oil-producing countries to work together because each distrusted the other. The Shah of Iran had a bitter example to remember. The Arab countries had increased their production to take up the slack when Iran faced an English boycott of Iranian oil in the 1950's.

At this point the oil countries finally realized that it was absolutely necessary for them to work together to raise oil prices. They had a clear American model for an oil monopoly in John D. Rockefeller, who had built his enormous empire by seeking control of the oil business. The Rockefeller monopoly was finally broken by antitrust laws passed by the United States Congress, but there was no international law body to tell the oil-producing countries they could not form a monopoly to get what they considered their share of the oil business.

The result of this dismay over dropping prices resulted in the formation of OPEC—one of the most effective monopolies in modern history. OPEC means Organization of Petroleum Exporting Countries. It was founded in September 1960 for the express purpose of raising posted prices of petroleum back to their pre-1960 levels and to bring about greater cooperation between oil-producing countries.

OPEC came about as the result of a meeting called by the Iraqi government in 1960. In addition to Iraq, oil ministers or representatives from Iran, Saudi Arabia, Kuwait, and Venezuela attended the meeting in Baghdad. They agreed that a strong organization of oil-producing countries was imperative. They estimated at the time that dropping oil prices and currency devaluations, coupled with inflation, had cost them collectively more than $600 million and that the situation would get worse unless they did something to assert their rights.

Plans were made for OPEC, which would have a permanent headquarters in Geneva, Switzerland (shifted to Vienna, Austria, in 1965). There was to be a permanent secretary and staff, including economists. Members would hold meetings twice a year, revolving the locations among their own capitals. Oil ministers would hold meetings every three months, and committees would meet whenever necessary.

In 1962, soon after the organization began functioning, Qatar, Indonesia, and Libya joined OPEC. They were joined the next year by Abu Dhabi, Nigeria, Algeria, and Ecuador. The present twelve-nation OPEC controls 93 percent of the crude oil exported outside Communist-controlled countries. OPEC also controls 75 percent of the world's known oil reserves. This gives them the absolute power to set world oil prices as they wish—as long as they can agree among themselves.

And they did agree well enough to quadruple the price of oil in two years, bringing economic hardship to a large portion of the world.

This enormous power over the world's oil did not come immediately. It took ten years for OPEC to become the strong, efficient cartel that it is today. Although formed from the 1960 Baghdad meeting, OPEC did not become a fully operating organization until 1963. At this time the members first put OPEC forward to bargain for them all. Naturally the oil companies resisted OPEC's demands, but in 1965 they gave in to the first one—that income taxes paid by the companies to the various governments could no longer be deducted as operating expenses in figuring royalties. It was estimated by the companies at the time that this would increase the governments' income by as much as nine cents a barrel of oil.

What is more important, OPEC formed a united front on posted oil prices, flatly refusing to permit any further reduction under threat of stopping all production. Thus the price of Arabian crude oil, which had dropped to $1.80 a barrel, remained at that price through 1970.

OPEC, as its membership showed, was not restricted to the Middle East. Its membership included countries in South America, North Africa, and the Orient. The Arab countries decided that while OPEC membership was necessary in order to form a strong cartel, OPEC did not fully serve their own intentions to use oil as a political weapon, especially against nations supporting Israel. They therefore formed, as a direct result of the 1967 Arab-Israeli War, a new organization which they called OAPEC—Organization of Arab Petroleum Exporting Countries.

OAPEC is often confused in the public mind with OPEC. They are not the same, nor do they have the same objectives. It was the OAPEC (Arab) countries which were involved in the 1973 oil boycott against the United States for assisting Israel.

In 1965 Libya, an Arab oil-producing state in North Africa, launched an attack on the oil companies' posted-price policy. The Libyan government announced that it and not the oil companies would determine future posted prices. Negotiations over this demand continued until June 1967, when the Six Day War between Israel and Egypt and Syria was fought. Negotiations were resumed later, with the oil companies standing firm for their right to determine posted prices. The Libyan government lost patience, issuing an ultimum with a September 1, 1969, deadline. The other Middle East oil-producing countries watched the negotiations closely.

There was a revolution in Libya before the deadline expired, and the oil companies hoped that it would help them. But the new military regime was as tough as the old government had been. The companies bowed to the inevitable. Libya then announced a 30-cents-a-barrel increase in the posted price of Libyan oil, effective September 1, 1970.

This broke the barriers all through the Middle East. Oil companies buying oil in the Persian Gulf did not wait for a demand

Special docks permit pipelines to carry oil directly to tankers at the Port of Aden on the southern shore of the Arabian Peninsula.

Courtesy British Petroleum Company

there. They automatically raised the posted price of Kirkuk and
Persian Gulf oil by 20 cents a barrel.

This move was too little and too late. The OPEC countries did
not intend to settle for less than the Libyans. So at the Baghdad
OPEC conference in November, 1969, the delegates approved a
resolution recommending that "member countries should seek to
ensure that posted prices and tax reference prices of their petroleum exports are consistent with each other."

They intended to use this resolution as a wedge for a gradual
increase in posted prices, but they were unable to do so because
there was still a surplus of oil. The Libyans were able to force
through their posted-price rise because two things worked in their
favor. One was the 1967 Israeli War. The Egyptians sank several
ships in the Suez Canal, closing the shortcut waterway to tanker
travel. Tankers then had to sail around Africa to reach the European market. This did not just raise the cost of transportation;
there was also a shortage of tankers to make the long haul. Since
Europe got 25 percent of its oil from Libya, a Libyan threat to cut
production was a serious matter to Europe.

Even so, the Libyan-based oil companies did not immediately
give in to the demands. However, the situation became extremely
serious in May, 1970, when an accident to the Tapline pipeline
carrying Saudi Arabian oil to the Mediterranean port of Zahrani,
Lebanon, was broken in Syria. The Syrian government refused to
let Aramco repair the line until the fee, paid for permission to
cross Syrian territory, was increased.

This stopped the flow of 480,000 barrels of oil a day to Europe.
Libyan officials immediately ordered a 600,000-barrel-a-day cut
in their production. The oil minister blandly reported that the cut
was not intended as a threat to Europe or as pressure on the oil
companies to agree to Libyan terms. It was, he claimed, a conservation measure to keep Libya itself from running short of oil.

Conservation was conveniently forgotten when the oil companies agreed to Libya's terms.

Immediately following Libya's success in forcing an upward

revision of posted prices, OPEC members met in Tehran in February 1971 and demanded the establishment of a 55 percent income tax rate on oil exports, a general increase in all posted prices, and the elimination of special discounts on oil. They also said they would demand price adjustments in line with currency-rate changes. The oil companies agreed to all these demands.

Then in January 1972, at the OPEC meeting in Geneva, the members demanded and got a 20-cents-a-barrel increase in price to offset devaluation of the United States dollar. The price was raised another 15 cents per barrel when the dollar was again devalued in February 1973.

The price of oil was steadily increasing, but not fast enough to suit the oil-producing countries. There is absolutely no denying that they got the short end of the stick through the later 1950's and the 1960's, when the price of their oil fell far behind the inflation rate that ballooned the cost of their imports. In all justice, they certainly deserved raises. The major point is whether they were justified in the enormous increases that were still to come— increases that have presented tremendous financial burdens to many small developing countries, and which plunged the world into an economic recession that could eventually boomerang back on the oil-producing states themselves.

The worst, however, was yet to come. A new and staggering increase came in October 1973, as a direct result of the Israeli-Arab war. A rise would have come even if the war had not broken out, for Iraq and Kuwait both had insisted in August that inflation had eroded the gains made by the oil-producing countries in boosting prices at the Tehran conference the previous February. However, no one was prepared for the shock when it came.

The Arab-Israeli war of 1967 ended in a smashing Israeli victory, with Israel occupying large sections of former Arab lands. One of the reasons for Arab losses was poor coordination between the different Arab countries and a lack of really modern equipment.

In the years following the 1967 war, Anwar al-Sadat, the

Egyptian leader, worked hard to bring about unity among the Arabs with the aim of regaining lost Arab land, especially the Sinai Peninsula, occupied by Israeli forces since 1967. Sadat was extraordinarily successful. He not only lined up support in Syria and Jordan, two Arab countries adjoining Israel, but got pledges of support from Iraq, Lebanon, Kuwait, and Saudi Arabia in the Middle East and from Sudan, Morocco, and Tunisia in North Africa.

The Arab attack began on Yom Kippur, the most holy Jewish festival, in October 1973. It was a slashing, two-front attack, with Egypt crossing the Suez Canal and driving into the Israeli Bar-Lev line in the Sinai. At the same time Syria attacked the Golan Heights, Syrian territory occupied by Israel since the 1967 Six Day War.

The attack was a surprise to the Israeli army, which fell back under the fury of the Arab assault. At the same time Iraq, Tunisia, and Kuwait sent troops to support the Egyptians in Sinai. Saudi Arabia also sent a token force, and Lebanon allowed Palestinian guerrillas to operate from its frontier to attack Israel along the Lebanese-Israeli border.

The war began on October 6. Israel regained the initiative by October 15 when its army crossed the Suez Canal, splitting two Egyptian armies. At the same time Israeli forces slashed into Syria to within twenty-five miles of Damascus, the Syrian capital.

Under United Nations pressure the combatants agreed to a cease-fire that went into effect on November 11. By this time neither side was in much condition to fight. Casualties had been heavy. Also, the loss of matériel and equipment had been enormous. Military experts estimated that the two sides lost as many as 1,800 tanks in what they described as the bloodiest tank battle in history. More than 200 airplanes were destroyed, mostly from increasingly efficient ground-rocket fire.

Both Russia, supporting the Arab side, and the United States, supporting Israel, mounted airlifts to resupply the combatants.

Oil, which had been a paramount military weapon since World

War I, now suddenly emerged as a political war weapon. As soon as the war began, the Arab nations, showing unusual solidarity, began a direct campaign on the political front. Abdul Rahman al-Atiqi, the Kuwaiti oil minister, called a meeting of OPEC members for October 16. The members almost immediately announced that they had agreed to a 70 percent increase in posted oil prices.

The day following the enormous 70 percent increase in prices, members of OAPEC met in Kuwait. Iran, which is not an Arab country, although it shares the Muslim religion, was not represented at the meeting.

At this meeting the decision was made to use Arab oil as a war weapon against the United States. There would be a total embargo of oil to the United States and 25 percent cutback in all Arab oil production so that oil going to other countries could not be resold to American markets. The plan had earlier been discussed between the oil ministers of Egypt and Saudi Arabia. The original plan called for a gradual cutback of 5 percent of production each month, but the plan was changed after President Richard M. Nixon announced on October 18 that he would send advanced weapons to Israel and would ask Congress to approve a $2.2 billion assistance grant to the Israelis. The Arab bloc then retaliated with a complete embargo of oil shipments to the United States and the Netherlands, both of which supported Israel in the war.

The effect was felt immediately in the United States. Lines piled up at service stations, which were forced by law in many states to close on Sundays. Gasoline was rationed to ten gallons at each filling, and there were threats of fuel-oil shortages for the coming winter. Congress passed emergency laws setting a nationwide 55-mile-per-hour speed limit and extending Daylight Saving time through part of the winter.

The United States, which once had supplied the world with petroleum, was running out of oil. Reserves were being used up. Producing wells were running dry and it was costing more to pull oil from those that were still producing. In November 1973, when the Arab oil embargo went into effect, the United States was using about 17 million barrels of oil a day. Of that, 6 million barrels a day had to be imported.

About 2 million of this daily requirement of 6 million barrels of imported oil comes from Arab countries. Something over 1 million barrels is direct import. The rest comes from countries where the original Arabian crude is processed into refined products for the American market. Most of the United States petroleum imports come from Venezuela. These were not affected by the boycott.

However, Venezuela is an OPEC member. The United States economy was staggered by the OPEC price raises, which jumped the cost of a barrel of oil from about $3 on October 1, 1973, to $11.65 on January 1, 1974. This enormous increase in the price of a basis commodity used in every level of society and industry drastically increased inflation, cut jobs, and triggered a business recession.

Heavy as the effect was upon the United States, it was a near disaster for poor, undeveloped countries, which were hit by increased fuel bills they could not afford to pay. In the United States only the 6 million barrels of imported oil were affected by the price increase. Domestic oil, except for newly opened fields, was kept under price controls, so that about two thirds of United States oil did not share the world price rise. There have been intense efforts by the oil industry to get these price controls removed.

The Organization for Economic Cooperation and Development (OECD) claimed that the huge OPEC price increases would cost importing nations together about an additional $63 billion a year. The extra cost to the United States alone was estimated at an additional $17 billion. The United States could more easily absorb this huge increase in oil payments because a lot of it would return to America in the form of large purchases of military weapons by both Iran and Saudi Arabia.

The quadrupling of oil prices through 1974 brought an enormous inflow of dollars into the treasuries of Iran, the Arab countries, and North African oil-producing nations. Yet, except for Kuwait and Abu Dhabi, most of them are in a constant crunch for still more money. This is partly due to extremely expensive development programs and extravagant expenditures for modern military weapons, and partly from waste and poor planning. In 1974, for example, Iran overspent by $5 billion and had to borrow from the International Monetary Fund.

The oil embargo and the accompanying production cut of 25 percent hurt the OAPEC countries by reducing their overall income as the consumer nations cut back on purchases. The embargo and cutback ended in March 1974, which eased the energy shortage in the United States but did not help the price situation. In fact, it led directly to still more price rises. In December 1974, OPEC added another 4 percent increase, but agreed—in view of the extreme hardships suffered by many countries—to hold that price through 1975.

No one expected oil prices to remain stable. There would be a new price rise at the end of 1975. The question the anxious world awaited was: how much?

The OPEC members met in Vienna in late September 1975 to fix the new prices. Rumors published by the press hinted that crude prices would rise as much as 20 percent when the meeting closed.

There were good grounds for this prediction. The majority of the OPEC members were pushing for an increase of 15 to 20 percent. At one point the Iranian delegate—who never takes any stand without direct orders from the Shah—was reported to be pushing for as much as 40 percent.

The success of the oil cartel in pushing up the price of oil was due solely to their united stand. Therefore, when evidence of near violent disagreement among them leaked out, the oil-consuming nations, hoped that some relief was in sight. If OPEC broke up, then competition would force prices down.

Jamshid Amouzegar of Iran angrily demanded that the stiffest

Middle East oil has made Mohammed Reza Shah Pahlavi, Shah of Iran, the most powerful man in the Middle East.

possible price be approved. Saudi Arabia's Sheikh Ahmed Zaki Yamani warned that a stiff increase could worsen the world's economic recession. Another member said contemptuously that it was not OPEC's job to solve the world's economic problems. It was their job to get the best possible return for their petroleum.

Although the last two rounds of price increases had raised the cost of Middle East oil five times what it had been in 1972, Iran's delegate, again echoing the Shah, insisted that the rise in the price of oil did not account for more than 3 percent of the world's inflation rate. The rest of the enormous rise was due to the failure of politicians to manage their economies properly.

Yamani bluntly told the assembly that Saudi Arabia would not agree to any price hike above 5 percent.

At that point tempers flared. Iran, which had suffered a $5 billion deficit in its 1974 budget, due to the Shah's grandiose spending program, bitterly denounced Yamani and his 5 percent demand.

After the turbulent meeting Yamani told newsmen that he was standing pat on his demand for a rise no higher than 5 percent. "I feel that anything higher will worsen the world recession," he said. "Just as the present recession has cut the amount of oil we are selling, a worse recession could further reduce our income even though our prices are higher."

Others saw his actions as a political bid to make Saudi Arabia the hero of the hard-pressed oil-consuming nations. Iraq's oil minister, Yayeh Abdul Karim, told reporters, "Yamani speaks only for Yamani. He does not speak for OPEC!"

When the session reconvened, Yamani gave some ground. He agreed to go along with a 10 percent increase, but only if 5 percent went into effect immediately and the other 5 percent was postponed until after January 1976. When the Iraqi oil minister bitterly assailed him, Yamani stood firm. "You cannot force through a big price increase unless Saudi Arabia agrees" he told them. They understood what he meant. Saudi Arabia had the production and oil reserves to set whatever price its rulers decided. They did not have to stand with the rest of OPEC.

After throwing this challenge at his fellow members, Yamani left the conference building. He told reporters that he was flying to London to telephone his government for instructions. He did not explain why he could not telephone from Vienna. Apparently the purpose of his London trip was to get away from his fellow oil ministers and not have to listen to their arguments until their anger at his challenge had time to cool and they could look at the matter more calmly.

There were no sessions while Yamani was gone. The oil ministers waited uneasily for his return. One of them told reporters that OPEC would be destroyed as a monopoly if Saudi Arabia failed to go along with the rest of them. None of them wanted that, not even Saudi Arabia. Tempers cooled and an agreement was made within a few hours—on Saudi Arabia's terms.

This halving of the expected 20 percent hike was hailed as a victory for the industrialized world. It was a victory of a sort, but it was still another severe blow to a world in recession. In the United States economists estimated that the 10 percent rise would push consumer prices up appreciably and add to unemployment. Gasoline prices would go up one cent.

The oil ministers met again in Bali in June 1976, after an earlier meeting in Vienna had been interrupted by terrorists. In view of the unstable world situation the ministers voted to continue a price freeze at the September 1975 level.

This was done with a very close eye on world economy. The drastic fall in world oil consumption had shocked their own economies, serving a warning that there is a point in price increases they cannot exceed without hurting themselves.

However, the oil-producing countries in OPEC do want their share and perhaps a little more. Price freezes will last only as long as they serve OPEC interests. Only Saudi Arabia and Kuwait among them have a real surplus of money. The rest are hungry for more. They can be depended upon to keep constantly pressing for more price raises any time their economists feel this can be done to an advantage.

12
Tomorrow's Energy

There is no reliable way to judge the future except by the past. Many deny the adage "History repeats itself." Mao Tse-tung, Founder of the People's Republic of China, was fond of saying, "History never repeats itself. When it tries, the result is a farce."

Yet history does try to repeat itself. Sometimes it succeeds; sometimes it fails. And when it fails it is usually because wise men have studied the past and taken proper steps to avoid the mistakes of yesterday.

In looking back over the history of petroleum we see from the very first a determined attempt by producers to establish monopolies and to set the highest possible price. It began in the United States with the Rockefeller interests and it continues today in the Middle East with the oil producers' solidarity.

Another thing we see in the past is American domination of petroleum production and sales. Today the United States' internal production is dropping and the cost of production is rising. Petroleum is not a product that renews itself. We can safely predict that the Middle East will lose its dominant position in petroleum, but just when we can only guess. Petroleum experts are estimating it

will happen about the year 2000. However, nobody really knows
the extent of Arabian and Iranian oil reserves. Over the past
twenty years the estimates on "proven" reserves have doubled as
new fields were brought in and old fields proved larger and more
productive than anyone realized. In addition, new production
methods permit oil companies to extract more oil from a given field
than was possible in the past.

The good life that we experience today in the Western world is
based upon widespread world industry that has been powered by
cheap energy. First the fuel for this energy was wood, then coal,
and finally petroleum in the form of natural gas and oil. Cheap
energy made mass production possible. Mass production in turn
brought costs down to the point where almost everybody could
afford automobiles, refrigerators, and other articles of necessity or
desire.

The capitalists tell us this good life is the result of the profit
system. The socialists and Communists tell us that if the profit
factor were removed and governments controlled production,
there would be still more for the people. Regardless of who is
right in this argument, it is an absolute truth than none of the
various "isms" can provide a decent standard of living for the
people *without an abundance of cheap energy*.

We have seen since 1973 what happens when energy costs rise.
There was a world recession. Shortages drove prices up at the
same time that inflation cut buying power. The sale of single-
dwelling houses dropped alarmingly because the average young
family could not afford the exorbitant down payment and high
interest rates. This in turn added to more unemployment.

In the United States we are asked to do with less—cut back on
house heating, turn out some of our lights, buy smaller cars, drive
at slower speeds, and pay out more of our inflation-riddled
salaries for everything we buy. At the same time we are hit with
higher taxes for welfare programs and unemployment.

Even so, the United States, which must import only 16 percent
of its energy needs, is better off than small developing countries,

which stagger under the load of increased energy costs. The only way these countries can gain a proper living standard is to develop industries, but the fivefold increase in oil prices has made such a goal impossible for many of them.

Some economists tell us that the days of cheap energy are over. They say we must brace ourselves to live in a future with a lower standard of living and possibly a high permanent unemployment rate. Other experts counter this prediction by insisting that alternate sources of cheap energy can be developed.

Many things have been suggested to take the place of Middle East oil. Solar energy, tidal energy, nuclear energy, thermal energy, a return to coal, and the development of America's enormous shale-oil reserves have all been suggested as future means of breaking the Middle East oil producers' stranglehold on the world's fossil fuel supply.

If there are so many alternate sources of energy, why haven't they been developed in the past and why aren't some companies working on them now?

They were not developed in the past because it takes a tremendous amount of money to develop a new industry, as oil producers found out when they were building their oil empires. Any new process is generally more expensive than an old, established industry. Alternate methods of energy in the past simply could not compete with cheap oil. In the 1960's, when petroleum was selling for $3.50 a barrel, oil cost 17 cents a barrel to produce in the Middle East. To this, of course, had to be added the cost of transportation and the cost of marketing. Even so, oil and natural gas were the cheapest known source of mass energy outside of hydroelectric power. But hydroelectric power, produced by dams, is limited, because just about all the places where giant dams can be built have already been developed.

There is a lot of experimentation in the commercial possibilities of solar energy. NASA (the National Aeronautics and Space Administration) has several programs under way. There are two basic ways to turn the sun's heat into solar energy to heat homes and for industrial use. In the direct method, the sun's rays are concentrated on water pipes and tanks by means of mirrors or lenses. The concentrated heat turns the water to steam, which then operates steam engines or generators. This method is not yet practical. It works only on sunny days. In addition, the lens systems require some means of constant movement to keep them in focus with the sun.

Some years ago the author saw a small solar stove for campers demonstrated by a Japanese inventor. It was made like an umbrella. It opened up and the silvered interior acted like a concave mirror to focus the reflected heat on a rice pot suspended in the center. It cooked rice very well. I ate some. But as with all such devices, when the sun went behind a cloud, you went hungry.

Photoelectric cells are receiving quite a lot of attention. These selenium cells generate small quantities of electricity when light shines upon them. The most familiar use of such cells is in exposure meters used by photographers. An advantage of such cells over direct-sunlight methods of solar energy is that photoelectric cells will work in diminished light as well as in direct sunlight, although the output is lower. However, they will not fail on a cloudy day as the direct-sunlight methods will. Because the amount of electricity generated per cell is small, hundreds of these cells are wired together to provide more power. Unfortunately, the experiments have been on a small scale because we do not yet have the technology to develop a cheap, practical solar-cell program.

There is a lot of interest in the possibility of solar energy, especially among ecologists who feel that it is safer and cleaner than any of the alternate energy sources. The Arthur D. Little Company, a group of research scientists, reported that the world could save 100 million barrels of oil each year if solar energy could be made to provide only one percent of the nation's energy needs.

A practical, safe, and clean means of generating energy is through geothermal processes. The word *geothermal* comes from a combination of *geo* (earth) and *thermal* (heat). This process uses the heat inside the earth for energy. In many places in the world steam gushes from the earth in the form of geysers, hot springs, and volcanic action. In Iceland, for example, hot water from geothermal springs is used for heating homes and even for hothouses to grow vegetables. Geysers in northern California are being successfully used for geothermal energy, and power plants in New Zealand and Italy are producing electricity from geothermal power. Where natural steam is not available, geothermal heat can be used by drilling geothermal wells in areas where underground heat can be tapped in an economical manner. The National Science Foundation reported in 1972 that 26,000 productive geothermal wells could produce 20 percent of the nation's electric power needs.

Nuclear power to generate electricity has been called the most promising alternative to petroleum. Some very successful nuclear-power plants are in operation today. Thirty-nine are in the United States, and sixty-nine are operating in foreign countries. In use, atomic fuel—uranium 235 enriched with uranium 238—creates heat when it reaches a certain mass. This fuel is suspended in zirconium cases and water circulates around the heated rods. The water boils from the atomic heat. The steam then runs a turbine, which powers electrical generators.

Uranium 235 is the material also used in atomic bombs. If the amount of U-235 reaches what is called "a critical mass," it will explode, with devastating results. Many environmentalists claim that nuclear-reactor power plants pose a danger to people. The Atomic Energy Commission has denied this, but even so, ecologists have successfully blocked the building of new nuclear-power plants.

Nuclear-power plants need large amounts of water to cool the reactors. This means that at the present time they can be built only in coastal locations or along rivers, which has led ecologists to ask courts to ban new plants because of "thermal pollution."

The United States has tremendous coal reserves—some 456

billion tons, according to estimates. However, coal is a dirty fuel. Soot and carbon can be removed from the smoke, but sulfur is much more difficult to remove. Strict environmental laws have discouraged the use of more coal at the same time that ecologists are attacking strip-mining. They claim that strip-mining, which lays bare large sections of earth, is destroying the environment. Even if coal could be produced as cheaply as petroleum, it is still more expensive to use. It has to be laboriously dug from the earth, loaded on trains, and shipped to market. Then it must be hauled again to the user, who must arrange for the coal to be fed into his boilers. It is inconvenient, bulky, and dirty.

No homeowner who has ever shoveled coal into a furnace and then carried out the cinders wants to give up an automatic oil or gas furnace to return to coal again. However, world coal reserves are about seven times as great as world oil reserves. So unless something is done soon to develop alternate energy sources, the world may be forced back to coal within the next twenty or thirty years.

In the opinion of some observers, the greatest future source of energy is from shale oil. Oil shale is a rock which, when heated, produces oil and gasoline. The United States has the world's largest reserves of oil shale, and 80 percent of it is on government-owned land in Wyoming, Colorado, and Utah.

No one really knows how shale was formed, but one theory claims that it is incomplete petroleum. That is, it was formed as they believe petroleum was, by thick layers of marine life and vegetation, which were then covered by sediments that became rock. Then the heat and pressure turned the layers of hydrocarbons into oil and gas. In the case of oil shale, they believe that there was not enough heat to complete the petroleum-building process.

In any event, if the oil-shale rock is mined, crushed, and heated properly, it will produce large quantities of petroleum. However, shale oil as it comes from the heated rock is very thick and filled with impurities. Either refineries have to be built right at the oil source, or the crude shale oil must be heated before it can be pumped through pipelines to refineries. The cost of mining shale, crushing it so it can be heated, and then subjecting it to temperatures of 850 to 900 degrees Fahrenheit, is expensive.

Practically nothing has been done to develop oil shale as an industry. The high cost has been given as the reason for the lack of development. Yet Marwan Iskandar, founder of Middle East Economic Consultants, claims that shale oil can be produced at costs ranging from $5 to $9 a barrel at the wellhead. This means that transportation and distribution costs and profits must be added. The price is well under the $10 to $12-plus that is being charged for Arabian and other Middle East oils.

Then why isn't shale oil being developed to free the United States from the iron grip of Middle East oil producers? The answer is that the petroleum industry is not interested in it. They have millions and millions of dollars invested in the present petroleum system and they do not want to spend billions more to develop shale oil until they are forced to do so by exhaustion of world oil reserves, which is not expected to occur until after the year 2000.

The quintupling of oil prices by Middle East oil producers did not affect the petroleum companies as it has the consumers. In the first year of the OPEC price squeeze, American oil companies reported amazing increases in revenue. One company reported to its shareholders that profits were up 200 percent. All the companies did was to pass along the enormous oil price increases to the public. Price controls were maintained on domestic oil production, but companies were free to add the extra price of foreign oil to their consumer prices. Profits have dropped since that record year, but are still about what they were before the oil crisis.

In justice to the oil companies, it must be pointed out that they are in business to make profits, and without profits they would be forced out of business. In order to attract investors to furnish capital to run and expand a business, industry must produce sufficient profits both to pay the investor for the use of his money and to provide extra capital for expansion. Oil companies claim that their present profit margins do not allow them this extra expansion capital and therefore they cannot adequately explore for more oil. There is considerable justice in their claims. We have seen the enormous amounts of money men like D'Arcy put into oil exploration. The original Iran wildcatting almost broke him, and the cost of drilling wells was much cheaper then than it is today.

It takes a minimum of a million dollars to drill the average well, and if it is an offshore well, then the price balloons alarmingly. One also must consider that a large number of wells drilled turn out to be dry holes—usually a total loss to the producer. Oil drillers today, as they have always been, are gamblers in a very expensive game.

We must not forget also that once oil is discovered, it takes five years or more before tank farms, pipelines, refineries, and other support can be built at a cost of extra millions of dollars in investors' money.

The ballooning cost of oil is not the only worry associated with Middle East oil. There is considerable concern about what the oil-producing nations will do with their oil money ("petrodollars," the newspapers call it). In 1974 alone the *extra* money rolling into Arab and Iranian coffers totaled about $15 billion. There is concern that these countries will use their enormous wealth to buy control of American companies. In that way they could get total economic control of the United States.

So far there have been some petrodollars invested in the United States, but not enough to be alarmed about. In fact, the United States has profited considerably from both Iranian and Arab military purchases in the United States. Since Saudi Arabia and other Middle East countries have expropriated and nationalized so much of Western-owned industries in their countries, they seem afraid to invest too much of their oil wealth in foreign nations for fear they in turn could lose it to nationalization.

This was stated clearly by Farouk M. Akhdar, an important

official in the Saudi Arabian planning agency, who said that the Saudis did not believe their money would be safe in the United States.

Arab suspicion of the United States was increased by a statement of President Gerald Ford that he might consider military action if the Middle East oil-producing nations attempted to "strangle" the United States economically.

The Shah of Iran reacted strongly to this implied threat. In an interview with Kingsbury Smith of the Hearst Press, the Shah said that such intervention would be illegal under the United Nations charter and also unnecessary. "The United States will never be strangled [for oil] because you have oil yourself," the Shah said. The Iranian ruler went on to say, "If there is another oil embargo we will not join in. We did not join in the last one, and we won't join in the next one. Oil is one thing. Politics is another. We don't mix oil with politics in that sense. So we would be out of the military intervention picture."

The Shah knows better than anyone the lengths powerful nations will go to protect their oil interests. He saw what happened in 1941 when Russia and Great Britain invaded his country and drove his father from the throne primarily to secure the Iranian oil fields. He saw it happen again at the end of World War II when Russia tried to continue occupation of the potential oil fields in northern Iran.

In the meantime, inflation and huge spending programs by the oil-producing nations make them even hungrier for money. Despite the enormous sums he is receiving, the Shah of Iran spent $5 billion more in 1974 than his country took in. Economists estimate that Saudi Arabia and Kuwait are the only two Middle East countries amassing huge cash surpluses. The big price increases by which the OPEC countries hoped to increase their earnings were partially offset by oil conservation in the using countries. Oil production dropped 11.3 percent in 1975 as compared with 1974.

There are still other worries ahead for the OPEC oil cartel. In 1975 Britain began receiving oil from its newly developed North

Sea oil fields. Estimates vary as to what this will mean. Some claim that the North Sea fields will not furnish more than 75 percent of Britain's needs. Other estimates claim that within five years Great Britain and Norway, which shares the North Sea fields, will be able to sell large quantities of oil to Europe. Within the same period an estimated 2 billion barrels of Alaskan North Slope oil will begin flowing into the United States, through the Alaska Pipeline, work on which was begun in 1974.

No one expects this to reduce the price of oil, but it will help reduce dependence upon Middle East oil and the power of vengeful Arab nations to declare embargoes for political reasons.

The only thing that can reverse the upward trend of energy costs is the development of an alternate energy source that is cheaper than petroleum. We keep hearing that the present high cost of oil has made other sources of energy competitive, but there is still no rush to develop these sources, and there will not be as long as oil companies can still make a profit on imported petroleum as they are doing today—and as long as these same companies are able to maintain their enormous domestic political contacts.

Politics and oil have been inseparable. Each has protected the other throughout the world. In recent years there has been some sign that the big oil companies in the United States are losing their political power as more and more liberal members of Congress are elected to office. One recent sign is congressional action to reduce the depletion allowance for oil companies. The depletion allowance permitted oil companies to deduct up to 20 percent of their income from federal taxes to make up for the fact that their oil reserves were being used up.

There are other setbacks for the oil companies as well. One is the continued regulation of prices on domestic crude oil and natural gas. This has continued in spite of attempts by President Ford to remove them.

Everyone realizes that Arab and Iranian domination of petroleum is only temporary. At best, present reserves will be used up in the next twenty-five years, experts say. But that does not take into consideration any future discoveries of new oil fields. Offshore drilling is increasing in the Persian Gulf, and new fields as important as those in the North Sea may be found in the future.

In the United States, offshore drilling on both the Pacific and Atlantic coasts could probably free the United States from dependence upon foreign oil. However, continued court actions by environmental groups have delayed drilling in these potentially rich areas.

Many of the Middle East countries are trying to develop industries to take up the slack when their oil runs out. Indications are that much of the money put into such development is being wasted. In contrast, one of the best examples of efficient use of oil billions is that of Abu Dhabi, the tiny oil-rich sheikhdom in the Trucial Coast region of the Arabian Peninsula.

Abhu Dhabi, led by the very enlightened Sheikh Raschid, who is also vice-president of the United Arab Emirates, is making wise use of its oil money. The country has an annual income of about $3.5 billion. One billion of this is spent in Abu Dhabi for the benefit of the people. Another billion is used for loans and grants to less fortunate Arab countries. The remaining billion and a half is invested in foreign countries. This investment takes the form of gold holdings, securities, and real estate.

Suggestions have been made that the other oil-rich countries should share their wealth with less fortunate countries, as Abu Dhabi has done. In their January 1976 meeting in Paris, OPEC members after long talks about aid finally agreed to give $800 million to developing countries hard hit by high oil prices. The original proposal was for $1 billion, but Indonesia and Ecuador refused to contribute and the others refused to make up the deficit. According to the Associated Press, the OPEC members would have total oil revenues for 1976 in excess of $110 billion, and their generosity to less fortunate nations would total less than three-fourths of one percent of this petroleum income.

In the meantime, the OPEC problem is how to get still more from the Western world in return for its oil. The Arab members can hardly be expected to show any mercy or friendship to those who buy their oil, for as they read history, from the time of the Crusades until today the Western world has given them little of either.

Although it is true that Saudi Arabia objected to the 20 percent price hike demanded by Iran at the December 1975 OPEC conference, this was because Oil Commissioner Yamani feared that too steep an increase would further erode oil sales and reduce profits in the long run.

In Iran particularly, Western cutbacks in oil purchases are creating a problem. The *Petroleum Economist,* an oil-industry magazine published in London, claimed that Iran's crude-oil production dropped 33 million tons in 1975 as a result of conservation among its European customers.

In interviews given in 1976 the Shah of Iran verified these figures and blamed the drop for the $4 billion deficit his country's finances suffered in 1975. He hinted that it would mean serious cutbacks in his country's purchases abroad. Such purchases, especially of military equipment, have been important factors in easing some of the financial impact of the oil price rises upon the United States.

The future looks bleak for both sides.

Important Dates in Middle East Oil History

150,000,000 B.C.	Petroleum formed in Middle East.
637 A.D.	Arab religious wars begin.
732	Arabs defeated at Battle of Tours.
1492	Muslims expelled from Spain by Ferdinand and Isabella.
1859	Edwin L. Drake drills first oil well in United States.
1862	John D. Rockefeller becomes interested in oil.
1882	Rockefeller forms Standard Oil trust.
1901	D'Arcy receives drilling concession for Persia.
1902	G. B. Reynolds begins drilling in Persia.
1908	Reynolds brings in first Persian oil well.
1909	British receive ground for refinery at Abadan.
1914	British government buys 51 percent interest in Anglo-Persian Oil Company.
1921	Reza Khan leads Persian revolt against Qajar government
1925	Reza Khan crowned as Reza Shah, king of Persia.
1927	Discovery well brought in Iraq's Kirkuk region.
1932	Oil discovered in Bahrain.
1933	Reza Shah annuls D'Arcy agreement and signs new agreement with Anglo-Persian.
1938	Discovery well brought in in Iraq's Kirkuk region.
1939	World War II begins, dampening and then expanding Middle East oil production.
1941	Iraq and Iran invaded by British and Russians.
1942	Abadan expanded; becomes world's largest refinery.
1948	Extraordinary expansion of Kuwait oil industry begins. Iraq closes pipeline to Palestine as protest to formation of nation of Israel.
1950	Tapline (Trans-Arabian Pipeline) opened.
1951	Mossadeq nationalizes Iranian Oil. British evacuate Abadan and Iranian oil fields. British embargo Iranian oil.
1953	Mossadeq proclaims republic; Shah flees, but returns after coup led by Iranian army assisted by CIA.
1954	Iranian fields reopened under operation of Consortium.

1960	OPEC formed at Baghdad meeting.
1965	OPEC establishes permanent headquarters in Vienna, Austria.
1967	Arab-Israeli Six Day War.
1969	Libya forces rise in posted oil prices.
1972	Iraq nationalizes Iraq Oil Company. OPEC shows gathering strength by forcing price rise to offset American dollar devaluations.
1973	Yom Kippur War. Arab oil producers place oil embargo on United States and the Netherlands.
1974	Embargo lifted. Britain opens North Sea field. Alaskan Pipeline work begins.
1975	New price rises bring Middle East oil prices to five times 1971 prices.

Bibliography

Arabian-American Oil Company, *Aramco Handbook: Oil and the Middle East*. Dhahran, Saudi Arabia: Aramco, 1968

Elwell-Sutton, Laurence P., *Persian Oil*. London: Lawrence and Wisehart, 1955

Kemp, Norman, *Abadan*. London: Allen Wingate, 1953

Longhurst, Henry, *Adventure in Oil. The Story of British Petroleum*. London: Sidgwick, and Jackson, 1959

Longrigg, Stephen H., *Oil in the Middle East: Its Discovery and Development*. London: Oxford University Press, 1968

Mohr, Anton, *The Oil War*. New York: Harcourt Brace and Co., 1926

Shwadran, Benjamin, *Middle East Oil and the Great Powers*. New York: John Wiley & Sons, 1973

Stegner, Wallace Earle, *Discovery:* Beirut, Lebanon: Middle East Export Press, 1971

Index

Abadan
  British evacuate, 114
  description of island, 34–35
  labor strikes at, 107, 108
  oil production in World War II, 95–96
  refinery, decision to build, 32, 35
Abu Dhabi, 87, 99, 153
Al-Bakr, Ahmad, 123
Al-Gailani, Iraqi nationalist, 91–92
Al-Wahab, Muhammad ibn, 76
Amouzegar, Jamshid, 139
Anglo-Iranian Oil Company, 74
  contempt to Iranian employees, 109
  rejected by Shah, 119
  strikes against, 107–108, 109
Anglo-Persian Oil Company
  British govt. buys stock, 40–41
  concession annulment, 66–68
  formed, 33
  name changed, 73
  royalty difficulties, 60–62
  Shah demands increased interest in, 55–56
Anticlines (geology), 52
Arab history, 8–9, 75–77
Arab-Israeli war, 135–136, 137
Arabian American Oil Co. (Aramco), 83, 89, 97
Arthur D. Little Co., energy survey, 146
Atatürk, Kemal, 59

Bahrain Island, 76, 79
  drilling on, 80, 81, 82
Bahrain Petroleum Co., 81, 82
Basra Petroleum Co., 122
Bechtel Corp., builds Tapline, 100
Bible, Wilson uses as code, 31
Bissell, George, oil pioneer, 13, 14
Boycotts, oil,
  Arabs against the Netherlands, 7
  Arabs against U.S., 7
  economic effects of, 8, 137–138
  political effects of, 8, 137
Burmah Oil Co., 18, 19, 27, 30

Cable-tool drilling, method, 22
Cadman, Sir John, 61, 65, 66
California Arabian Standard Oil Co.

(Casoc), 83–85, 87–89, 98
Casing, Drake invents, 14
Chiah Surkh, first Iran well, 20, 22, 26
Churchill, Winston S.
  appoints oil commission, 39
  First Lord of Admiralty, 39
  interest in oil, 27
  urges purchase of AIOC, 40–41
  wants invasion of Iran, 92–93
CIA activity in Iran, 117, 118
Clemenceau, Georges, writes Wilson, 47
Concession, effect of new, 70–71
Cox, Sir Percy, political officer
  arranges Abadan treaty, 31
  defines Kuwait border, 78
Crusades, the, 9
Curzon, George N., 43, 50, 90

Dammam Dome (oil field), 84, 87, 88
D'Arcy, William Knox
  aided by British government, 26
  financial difficulties of, 21, 26, 27
  gets Persian oil rights, 20
  hires driller for Persia, 20
  interest in oil, 17, 18
  reaction to oil strike, 31, 32
  sells interest to Burmah Oil, 26–27
Drake, Col. Edwin L., 14

Eastern and General Syndicate, 79
Edison, Thomas A., 18
Eisenhower, Dwight D., refuses Iran aid, 117
Embargo, oil; see boycott, oil
Energy, alternate sources
  direct sunlight, 147
  geothermal, 146–147
  nuclear, 147
  solar, 145–146
Exploration, oil
  in Arabia, 78–79
  by Casoc, 83–85, 87
  in Iraq, 91
  by seismograph, 87

Faisal, King of Iraq, 43, 48, 50, 91
Fisher, Sir John

157